AF730620

Photovoltaic Power Systems

For Inspectors, Plan Reviewers & PV Professionals

BASED ON THE 2017 *NATIONAL ELECTRICAL CODE*

About the author

John Wiles is perhaps the most recognized name in the solar industry for his numerous contributions to the development of codes and *National Electrical Code* compliance for photovoltaic systems. He has written hundreds of articles on *Code*-related photovoltaic system topics and is a regular contributor to *IAEI News*.

Wiles retired from his full-time position as a research engineer at the Southwest Technology Development Institute at New Mexico State University in 2013 after 24 years in the position. He continues to volunteer his time to keep active in the codes- and standards-development processes; to assist inspectors and plan reviewers with *Code* questions; to make "PV and the *National Electrical Code*" presentations throughout the country; and to consult with the PV industry on code-related issues.

Wiles is a member of several Standards Technical Panels for Underwriters Laboratories and is active in formulating standards for PV equipment, such as modules, inverters, charge controllers, combiners, cables, racks, and connectors. He continues to write the "Perspectives on PV" articles in the *IAEI News* and is active in the development of proposals for the 2020 *NEC*.

As an old solar pioneer, he lived for 16 years in a stand-alone, off-grid, PV-powered home in suburbia. His new owner-designed and -built retirement home has a 7.5 kW utility-interactive PV system with whole-house battery backup, where he lives with his wife Patti, three dogs, and two cats.

Photovoltaic Power Systems

For Inspectors, Plan Reviewers & PV Professionals

BASED ON THE 2017 *NATIONAL ELECTRICAL CODE*

THIRD EDITION

JOHN WILES

Published 2012, 2014, 2018 by
International Association of Electrical Inspectors
901 Waterfall Way, Suite 602
Richardson, TX 75080-7702

 First Edition published June 2012.
Printed in the United States of America
22 21 20 19 18 5 4 3 2 1

ISBN-13: 978-1-890659-81-3

Photos used in this book were shot in situ or at tradeshows. Use of the photos does not imply endorsement by IAEI of the manufacturers or the products.

The material in this book has been extracted from and expanded upon the series of articles "Perspectives on PV" found in *IAEI News* published by the International Association of Electrical Inspectors. The articles are based on the author's understanding of the 2005, 2008, 2011, 2014, and 2017 NFPA 70 *National Electrical Code (NEC)* [1]; his activities in developing that *Code*; his design reviews, inspections and testing of photovoltaic (PV) systems for more than twenty years; and his interaction with electrical inspectors, PV systems designers, and PV installers throughout the country. In all cases, the *NEC* is the requirement and local authorities having jurisdiction provide the interpretations of the *Code*.

DISCLAIMER

This book provides information on how the 2005, 2008, 2011, 2014, and 2017 *National Electrical Codes* apply to photovoltaic systems. The book is not intended to supplant or replace the *NEC*; it paraphrases the *NEC* where it pertains to photovoltaic systems and should be used with the full text of the *NEC*. Users of this book should be thoroughly familiar with the *NEC* and know the engineering principles and hazards associated with electrical and photovoltaic power systems. The information in this book is the best available at the time of publication and is believed to be technically accurate. Application of this information and results obtained are the responsibility of the user.

In most locations, all electrical wiring (including photovoltaic power systems) must be accomplished by, or under the supervision of a licensed electrician and then inspected by a designated local authority. Some municipalities have additional codes that supplement or replace the *NEC*. The local inspector has the final say on what is acceptable.

This book has not been processed in accordance with the National Fire Protection Association's (NFPA) *Regulations Governing Committee Projects*. Therefore, the text and commentary in it shall not be considered the official position of the NFPA or any of its committees and shall not be considered to be nor relied upon as a formal interpretation of the meaning or intent of any specific provision or provisions of the 2005, 2008, 2011, 2014, or 2017 editions of *National Electrical Code*.

Author and the publisher do not warrant or guarantee any of the products described herein nor have they per-formed any independent analysis in connection with any of the product information contained herein. The publisher does not assume, and expressly disclaims, any obligation to obtain and include information referenced in this work.

The reader is expressly warned to consider carefully and adopt all safety precautions that might be indicated by the activities described herein and to avoid all potential hazards. By following the instructions contained herein, the reader willingly assumes all risks in connection with such instructions.

THE AUTHOR AND THE PUBLISHER MAKE NO REPRESENTATIONS OR WARRANTIES OF ANY KIND, INCLUDING, BUT NOT LIMITED TO, THE IMPLIED WARRANTIES OF FITNESS FOR PARTICULAR PURPOSE, MERCHANTABILITY, OR NON-INFRINGEMENT, NOR ARE ANY SUCH REPRESENTATIONS IMPLIED WITH RESPECT TO SUCH MATERIAL. THE AUTHOR AND THE PUBLISHER SHALL NOT BE LIABLE FOR ANY SPECIAL, INCIDENTAL, CONSEQUENTIAL OR EXEMPLARY DAMAGES RESULTING, IN WHOLE OR IN PART, FROM THE READER'S USE OF OR RELIANCE UPON THIS MATERIAL.

1 *National Electrical Code* and *NEC* are registered trademarks of the National Fire Protection Association, Inc., Quincy, MA 02169

Table of Contents

Preface

In a time where photovoltaic plan reviewers and inspectors are getting pressured to expedite the inspection and review process, this book could not have come at a better time. If we are expected to accomplish faster turnaround times with fewer inspections, we must be informed to ensure safe, *Code*-compliant installations. Rubber-stamping plans and drive-by inspections may be what the industry is pushing for, but are those actions what the customer deserves? The customer is relying on a qualified inspector to verify that their PV system is safe and that it will continue to be safe for years of operation.

What makes this book stand out is how it correlates to the *National Electrical Code* (*NEC*). When citing corrections or comments, we need to be able to reference the *Code* to justify our calls. The last thing that we should be doing is trying to enforce our opinion.

Stamping a set of PV drawings for approval or signing a permit card for an inspection does not require skill or knowledge of the *NEC* requirements we need. The knowledge and skill before we sign or stamp documents is where most of us need some help and guidance. Having a document such as *Photovoltaic Power Systems* provides inspectors with a great tool for gathering information on what to look for in plan review and during inspections.

Article 690 is a small section when compared to the entire *NEC*. The size of Article 690 does not make it any less important than other articles found in the *NEC*. Due to its size, it is often not a focus of the combination inspector. It is no wonder it gets overlooked when you stack up all the codes the combination inspector must enforce.

When the *NEC* book and handbook are not enough to help you understand what or how you should be enforcing the regulations, this document can provide clarity. The information provided in *Photovoltaic Power Systems* has been compiled by someone who is known throughout the industry as a PV expert. John Wiles has been a resource and has been providing training for more than 20 years to inspectors and plan reviewers.

Not only is this book originally based on the 2011 *Code* and earlier editions, but also the 2014 and 2017 *NEC*. With knowledge comes credibility. This document will help plan reviewers and inspectors know and understand what they are looking at and what to look for. If you want to understand what is on the plans, this book will help. Even for those who have a solid understanding of PV systems, it is helpful to have a book to refer to when questions arise. Whether questions come from us or from PV designers or installers, this book can help answer these PV-related questions. You will find this book to be an excellent resource.

Having ready access to this book can help us all be more informed about PV systems, where expertise is often limited. John Wiles is known for having such expertise.

— Rhonda Parkhurst, Electrical Specialist
City of Palo Alto, California

01

An Overview of PV Systems and the 2017 ***National Electrical Code***

Photovoltaic (PV) power systems are being installed by the tens of thousands throughout the United States. In states where financial incentives are available (like in California, New York, and New Jersey), the PV business is booming. The first PV cells produced more than 50 years ago are still producing power, and modern PV modules are expected to produce energy for the next 40 years or longer. The power output from PV systems ranges from a few hundred watts to many megawatts. Most of the systems are *not* operated or owned by any electric utility and therefore come under the requirements of NFPA 70 *National Electrical Code* (*NEC*). Unless otherwise noted, all references to the *NEC* will be to the 2017 *NEC*.

Systems as large as 700 megawatts have been installed by third parties on private land in the United States, and are not under utility control or ownership. Larger systems of up to 1500 megawatts have been installed in other countries and will more than likely be installed in the United States in coming years. These systems operate at 1000 volts to 1500 volts and, in the larger commercial systems, direct current (dc) and alternating current (ac) can range up to 2000 amps or more. These levels of voltage and current, if not properly managed, pose shock, life safety, and fire hazards. These systems must be inspected to ensure the safety of owners, operators, service personnel, and the public.

The *Code* requirements for a typical residential PV system are at least as complex as those for residential wiring, and the dc portions of the system coupled with the ac interconnection to the utility grid make PV installations unique. Because the PV industry is growing rapidly, individuals, companies, and organizations (with varying degrees of knowledge, skill, and experience) are installing these systems. Large, and some small, PV-system integrators and vendors — working with experienced electrical contractors who have jointly pursued additional PV-specific training and who work closely with the local permitting and inspecting authorities — usually (but not always) perform the best, most *Code*-compliant installations.

On the other hand, individuals or organizations with little or no experience or training installing electrical systems of any type are installing many new PV systems. These systems may be unsafe (not *Code*-compliant) at initial installation;

develop hazardous conditions over the life of the system; be hazardous to operate or service; and fail to deliver the full performance of a well-designed and -installed PV system.

The electrical inspector or plan reviewer, as the authority having jurisdiction (AHJ), is the key player in ensuring that these less-than-ideal PV installations do not proliferate. Inspectors need to demand additional training in the inspection of PV systems and then inspect these systems very closely. Yes, PV is a relatively unfamiliar technology, but 80% of the *Code* already familiar to inspectors applies, and it is relatively easy to learn the inspection requirements that are unique to PV systems.

Several organizations in the PV industry provide training and certification for individuals. The training, experience, and skill requirements of PV designers and installers obtaining this certification will help ensure that safer, higher quality installations of PV systems take place.

PV System Types

Two main types of PV systems are being installed in the United States: utility-interactive (grid-connected) (see photo 1.1) and stand-alone (off grid) (photo 1.2). Both types use PV modules connected in series and in parallel to form PV arrays that produce direct current energy at voltages ranging from approximately 12 volts to 1500 volts (photos 1.1, 1.2, 1.3 and 1.4). Refer to Article 100 and Section 690.2 of the *NEC* for definitions of the terms used to describe PV equipment and systems. These systems will be

Photo 1.1 • Carport PV systems generate energy and keep cars cool.

Photo 1.2 • 3.3 kW stand-alone, off-grid PV system.

Photo 1.3 • Commercial PV array mounted horizontally with some shading.

Photo 1.4 • Five mW utility-scale PV system with PV subarrays mounted on two-axis trackers.

defined and further described in the following chapters

Generally, energy storage batteries are found in stand-alone systems but are not normally found in utility-interactive systems. Variations of each system are possible with some utility-interactive systems having battery banks to provide energy when the utility power is not available. These multimode PV systems are becoming more common as energy storage system prices decline, the utility power grid becomes less reliable, and natural disasters like hurricanes result in significant periods of utility power loss. Larger residential stand-alone systems will usually have a back-up generator, and these systems are known as hybrid stand-alone systems (photo 1.5).

Photo 1.5 • 10 kW backup generator, powered by natural gas, used when utility is out during extended periods of cloudy weather.

Photo 1.6 • Utility-interactive inverter. External ac and dc disconnects not shown.

Utility-Interactive Systems

Utility-interactive (U-I) PV systems are by far the most numerous type of PV system currently being installed. A typical residential system might have a PV array and an inverter (which converts dc to ac) capable of delivering 3000 watts to 10,000 watts of ac power to either ac loads in the house or to the utility grid when the PV power output is in excess of those local loads. In residential PV systems, single and multiple-inverter installations are common. The single inverter may have an ac output rating of 2000 watts to 7000 watts or more, and systems are frequently seen with two to four inverters used to increase the system power output (photo 1.6). A few residential PV systems have had ac outputs up to 90 kW!

These residential-sized inverters interface with the grid at 120 volts or 240 volts; are certified or listed to Underwriters Laboratories Standard for Safety 1741 (UL 1741 — *Inverters, Converters, Controllers and Interconnection System Equipment for Use With Distributed Energy Resources)*; and have all the necessary safety equipment built-in and verified as part of the listing process.

In commercial systems, the three-phase inverters commonly used usually start at approximately 10 kW and go up to 250 kW and to 2.5 MW (yes, 2.5 megawatt in a single inverter). They interface with the grid at 208 to 480 volts and higher (photo 1.7).

Stand-Alone Systems

Stand-alone systems are typically installed in remote areas where the utility grid is not available or where the connection fees to the grid are higher than the costs of an alternative energy system. While stand-alone systems sales are far lower than sales in the fast-growing utility-interactive PV system business, there is and has been a steady market for off-grid systems.

The stand-alone inverter converts dc energy

Photo 1.7 • 2.2-MW Sunny Central Inverter. *Courtesy of SMA Solar Technology AG.*

Photo 1.8 • Four 6-kW stand-alone inverters in an ac-coupled, battery-backed-up PV system

Photo 1.9 • Inverters and charge controllers for 10-kW off-grid, stand-alone PV system.

stored in batteries by the PV array to ac energy to support the loads (photos 1.8 and 1.9). Inverter power ratings are from about 250 watts to 6500 watts for residential systems and, as before, multiple inverters may be connected together for greater power outputs. Battery banks usually operate at a nominal 12, 24, or 48 volts, so the current levels to the inverters can be hundreds of amps at full load.

Larger stand-alone systems can be found in national parks, at telecom sites, and at federal facilities. These can be as small as residential systems with ac outputs in the 2 kW to 10 kW range, but they can also have single inverters of 250 kW or more. A few of these larger systems have multiple large inverters with combined outputs approaching 500 kW or more. Battery banks for the larger systems operate in the 200-volt to 600-volt range and dc currents to the inverters can be hundreds of amps at these higher voltages.

PV System Component Descriptions

PV Modules

The first thing inspectors see are PV modules. While most have glass fronts, aluminum frames (colored mill-finish aluminum or anodized brown or black), and plastic backs, some will be made with plastic frames or with no frames (photo 1.10). Others will be used as roofing materials (photo 1.11) or laminated directly to standing seam metal roofs (photo 1.12). PV modules come in many sizes and shapes.

Inspectors need to determine the listing of the modules and the electrical ratings. These are printed on the back of the module and may be available

Photo 1.10 • *(left)* Framed PV modules (anodized and clear-coated aluminum in natural aluminum color or brown/black).

Photo 1.11 • *(top right)* Building-integrated photovoltaics (BIPV) PV modules as roofing material.

Photo 1.12 • *(bottom right)* Thin-film PV modules laminated to a metal standing seam roof.

in the instruction manual or module specification sheet. Some unlisted, custom modules are being installed in architect-designed projects. Unlisted modules are being sold through various channels (including the internet), but unlisted modules no longer meet *Code* requirements and should not be installed [690.4(D)]. Although appearances may differ, these PV modules all produce electricity when illuminated and the normal cautions associated with any electrical power system should be followed.

PV modules come in differing power and voltage ratings and the sizes and ratings are continually changing. The modules must be connected in a manner that produces the needed voltage, current, and power because the output of a single module is usually not sufficient to operate the connected equipment or provide the needed amount of energy.

PV Combiners

PV combiners (PV j-boxes or PV combining enclosures) are common in PV systems operating at dc nominal voltages of 12, 24, and 48 volts and are also used in higher voltage systems (up to 1500 volts). They must be certified/listed by a nationally recognized testing laboratory (NRTL) to UL Standard 1741 [690.4(D)]. In these systems, it is a normal practice to connect modules in series (called a PV source circuit [690.2]) to get the proper voltage and then to connect each series source circuits of modules in parallel with other source circuits through a PV combiner to increase the current to get the desired power level.

These combiners will usually contain the overcurrent devices (fuses in the high voltage systems or circuit breakers in the 12-, 24-, or 48-volt systems) that are required to protect the module interconnecting conductors from fault currents

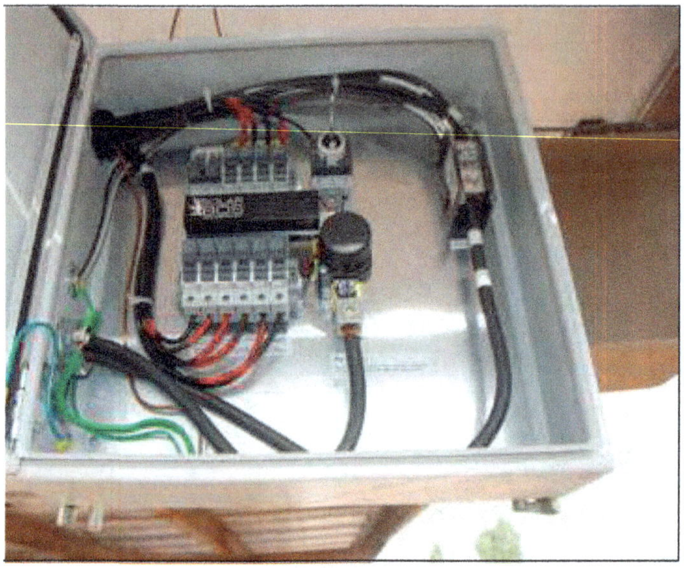

Photo 1.13 • PV combiner with fuses in the positive conductor only. Manual and contactor disconnects in positive conductor only. Not 2017 *NEC* compliant [690.15(C)].

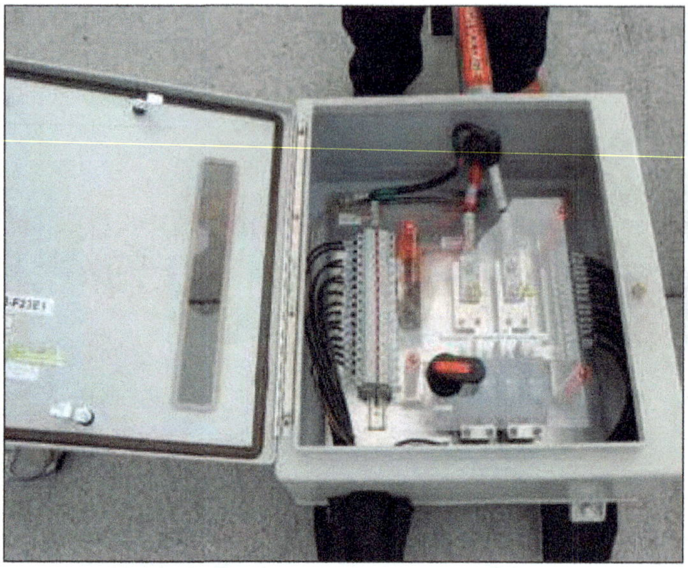

Photo 1.14 • PV combiner in white enclosure with manual switch opening both positive and negative conductors. Plastic shield covers energized, exposed conductors.

and the individual modules from reverse currents. Reverse currents may originate from parallel-connected strings of modules; reverse currents from the batteries in a system that has them; or from backfeed currents from a utility-interactive inverter (unlikely in listed inverters). See chapter 2 for additional details on the requirements for combiners.

Inverters

Inverters are found in both stand-alone systems and utility-interactive systems. They essentially convert dc)energy from the PV system (or the dc energy stored in batteries) to ac energy for use by local loads or for feeding into the utility system (photos 1.15 and 1.16). Some utility-interactive inverters, known as multimode inverters, have the capability to power selected load circuits from batteries or the PV system when the utility is not present.

Many PV owners in California were surprised when their utility-interactive PV systems did not work during the rolling utility blackouts created by the energy shortage and brownouts a few years ago. Utility-interactive PV systems with multi-mode inverters (photo 1.17) and battery backup were popular for months following the blackouts.

Unfortunately, installation manuals for these complex inverters (particularly the stand-alone types) can be several hundred pages long. The inspector should verify the proper dc and ac conductor sizes and overcurrent protection. Both are based on the rated ac power output of the inverter. (See Sections 690.8 and 690.9 and Articles 705 and 706 in the *NEC*.)

Utility-interactive inverters have all the automatic ac utility disconnect devices built-in, which protects utility linemen who are working on supposedly de-energized utility feeders. The utility-interactive PV inverter will not energize a dead line and, in fact, will disconnect from the line when the line voltage varies more than -12% to +10% from nominal (typically 120, 208, 240, 277, or 480 volts) or when the frequency varies by more than -0.7 to +0.5 Hz from the normal 60 Hz.

The inverter monitors the utility line voltage and frequency, and that voltage and frequency must remain stable and within tolerance for five

Photo 1.15 • Dual 4 kW utility-interactive inverters.

Photo 1.16 • Transformerless (non-isolated) inverters, typically used on most recent residential and small commercial PV systems. *Courtesy of SMA Technologies AG.*

minutes before the inverter can resume power transfer from the dc output of the PV system to the ac loads or to the utility.

All PV systems, with a few exceptions, must have a device known as a ground-fault detection/interruption device (GFID). *See* 690.41. These GFIDs are normally built into all utility-interactive inverters and some charge controllers. Early utility-interactive inverters used a fuse that blew on ground faults, but this method proved inadequate in detecting all ground fault, particularly those in grounded circuit conductors. The stand-alone inverter will usually not include the Section 690.41 ground-fault device, so if it is not built into the charge controller, an external, field-installed ground-fault protective device must be used. The author developed a system of circuit breakers in 1990 to meet the existing *NEC* requirement, but like the use of a fuse GFDI, it was not adequate to detect all low-level faults in grounded conductors.

Photovoltaic systems mounted on buildings will require a PV rapid shutdown system (PVRSS) which is system manually activated by first responders to reduce electric shock hazards due to energized PV conductors on or in a building (690.12). Many of these systems will be built into the U-I inverter, but separate versions are also available.

Photovoltaic systems with dc circuit voltages over 80 volts will require a device known as a dc PV arc-fault circuit interrupter (DCPVAFCI) and, in most cases, these devices will be built into the string inverter (690.11).

PV and the 2017 *National Electrical Code*

The 2017 *NEC* incorporated sweeping changes in the way it addresses requirements for PV and related systems and that will resonate into the future.

Changes Are Here

Two main areas stand out among the numerous changes that have been accomplished by Code-Making Panel 4 and the NFPA Technical Correlating Committee and its task groups. The first major change is a redefinition of what a PV system is and the exclusion or removal of most non-PV system information and requirements from Article 690. The second major change is a redefinition of what is meant by "grounding" and "grounding requirements" as they pertain to PV systems.

Photo 1.17 • Multimode Inverters. These inverters provide whole-house battery-backed-up energy and are ac coupled to utility-interactive inverters.

PV Systems Redefined

The first major change to Article 690 was to more narrowly define exactly what a PV system is. As we know them today, PV systems may include several interconnected sources of energy, including battery storage systems, multimode inverters, generators, and similar devices. Photovoltaic system disconnects are located in various locations throughout the "so-called" PV system in a manner that is sometimes confusing.

In the 2017 *NEC*, the PV system is primarily concerned with and defined by the PV modules and any device connected directly to them that does not involve energy storage or another source of energy. Changes to Diagram 690.1 will highlight these differences in the definition of a PV system. Primarily, we are looking at the PV system ending where the PV system disconnect occurs. In the 2017 *Code*, the PV system disconnect may be either an ac or dc disconnect depending on the configuration of the system. For example, as shown in Figure 1.1 (part of Figure 690.1 from the 2017 *NEC*) the PV system disconnect on a pure utility-interactive PV system occurs at the ac output of the utility-interactive inverter. If there are ac PV modules involved, the PV system disconnect will be at the output of the combined outputs of all the ac PV modules in the system. These PV disconnects separate the PV system from another energy source: the utility.

Even though it is acknowledged that the diagrams are not complete, the multimode dc-coupled system diagram (Figure 1.2) and the stand-alone system shown (Figure 1.3) are possibly confusing in one aspect. Typically, the output of a PV array cannot be connected directly to a stand-alone or multimode inverter, to dc loads, or to an energy storage system as shown in these diagrams. If done as shown, the PV output voltages to these circuits would be unregulated (changing throughout the day and as clouds appear) and

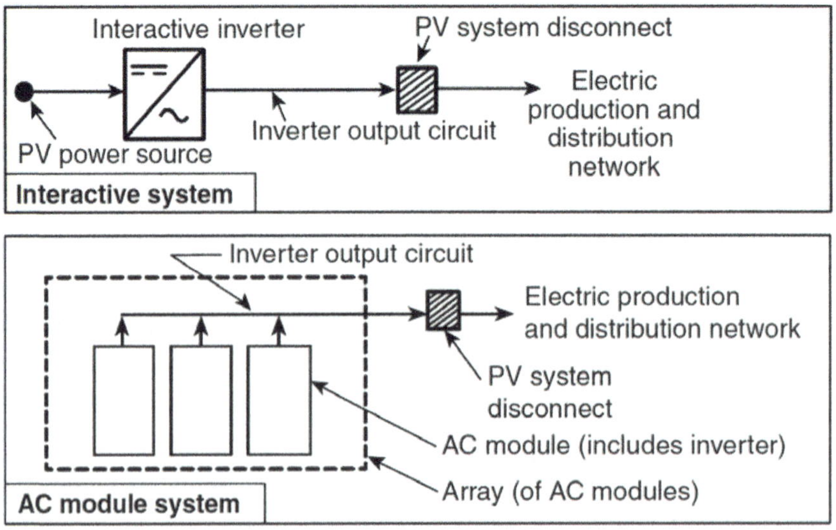

Figure 1.1 • Disconnects for simple PV systems. From Figure 690.1 in the 2017 *NEC*.

could possibly damage the connected equipment. Some sort of charge-management device (e.g., a charge controller) is normally used, and under the new definitions, that charge management device or system should be grouped with the energy storage system. If that is the case, then the PV system disconnect would properly be connected to an energy storage disconnect and connections to the inverter and any dc loads would be made directly to the voltage-regulated energy storage system or battery through appropriate disconnects. Also, in the two multimode diagrams (Figure 1.3), the output from the multimode inverter marked "Stand-alone system loads" in the dc-coupled multimode system diagram, and not identified at all in the ac-coupled multimode system, could probably be more properly identified as "Local ac loads" because they can be powered either by the utility or by the PV/energy storage system when there are utility outages.

As the PV system proper became more narrowly identified, parts of Article 690 that deal with other types of electrical systems have been moved to other parts of the *Code*.

Other Definitions Changed, Deleted, or Added

In Section 690.2, in addition to *Functional Grounded PV systems* (described below), other definitions were changed. The following definitions were also added.

Generating Capacity. The sum of parallel-connected inverter maximum continuous output power at 40°C in kilowatts.

Interactive Inverter Output Circuit. The conductors between the interactive inverter and the service equipment or another electrical power production and distribution network.

Photovoltaic System DC Circuit. Any dc conductor supplied by a PV power source, including PV source circuits, PV output circuits, dc-to-dc converter source circuits, or dc-to-dc converter output circuits.

The definition of ***Photovoltaic Systems Voltage*** was removed and now appears in Section 690.7.

Several definitions were modified, and they include:

Inverter Input Circuit. Conductors connected to the dc input of an inverter.

Inverter Output Circuit. Conductors connected to the ac output of an inverter

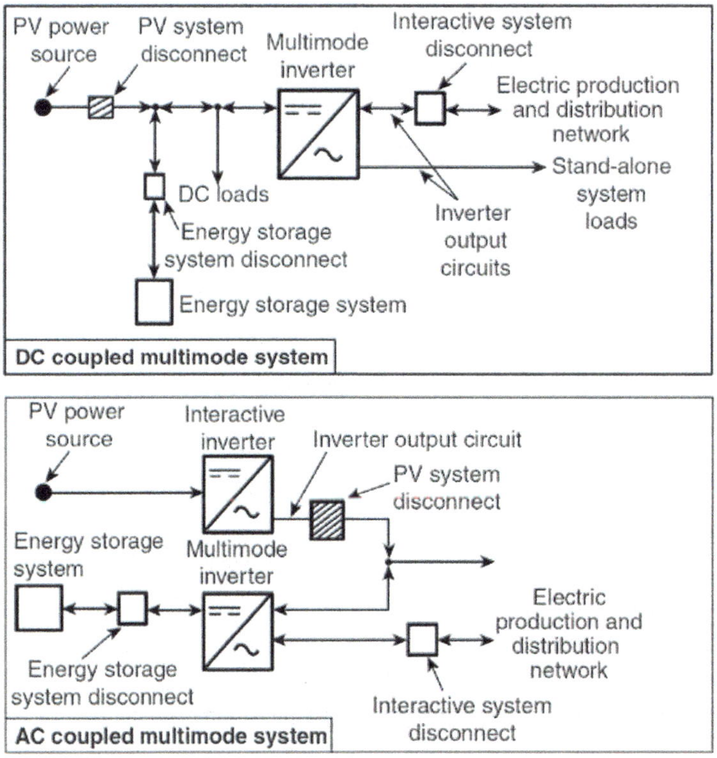

Figure 1.2 • Multimode, utility-interactive PV systems with energy storage (FIG 690.1).

Grounding—New and Old

Grounding PV systems have remained essentially unchanged for the most part since PV came to the *Code* in 1984. However, in the 2017 *NEC*, the terminology that is used to describe the various grounding functions has been changed.

This was done to reduce confusion by simplifying terminology between the older "Grounded PV Array/Isolated Inverter" PV systems and the newer, more common "Ungrounded PV array/ Non-isolated inverter" PV systems.

Grounding the PV System Circuit Conductors. In the early years of PV systems, up to about 2005, we primarily had grounded PV arrays and grounded dc battery systems where one of the circuit conductors was "grounded" or "connected" to the grounding system, which is composed of the equipment-grounding system, the grounding electrode conductor, and the grounding electrode system.

In some cases, the grounding method was a solid conductor and, in others, the grounding method was by a fuse or circuit breaker like those used in Section 690.5, Ground Fault Protection (2014 *NEC*). In other cases, various resistances or solid-state devices were used to ground one of the circuit conductors.

Now we have increasing numbers of ungrounded PV arrays and non-isolated (transformerless) inverters that have no dc circuit conductors connected directly to ground. Grounded PV systems, where one circuit conductor is solidly grounded, are becoming a thing of the past, but will exist for years and still be installed in smaller systems that are not on buildings. The required *Code* requirements to address these varying "grounding" systems were cumbersome, difficult to understand, and hard to apply in many cases. The changes in the 2017 *NEC* address and simplify many of these grounding requirements.

Grounding Definitions—2017 *NEC.*

From the 2017 *NEC*, some definitions are the same as in the 2014 *NEC*.

From Article 100

"**Ground.** The earth.

"**Grounded (Grounding).** Connected (connecting) to ground or to a conductive body that extends the ground connection.

"**Grounded, Solidly.** Connected to ground without inserting any resistor or impedance device."

Author's note: This definition could also include fuses, circuit breakers, and possibly relays and contactors with the devices not allowed in a solidly grounded system.

"**Grounded Conductor.** A system or circuit conductor that is intentionally grounded."

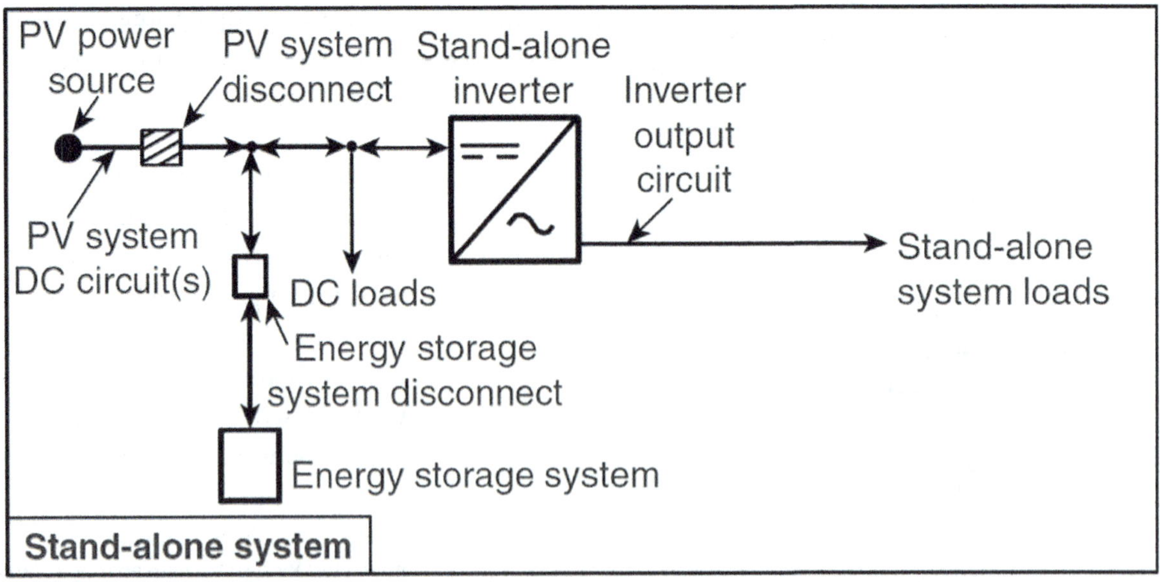

Figure 1.3 • Stand-alone system with energy storage (FIG 690.1).

From Article 690

> "**Functional Grounded PV System.** A PV system that has an electrical reference to ground that is not solidly grounded.
> **Informational Note:** A functional grounded PV system is often connected to ground through a fuse, circuit breaker, resistance device, non-isolated grounded ac circuit or electronic means that is part of a ground-fault protection system. Conductors in these systems that are normally at ground potential may have voltage to ground during fault conditions."

Author's notes: It is anticipated that revisions to the 2020 *NEC* will correct the grammatical error in the phrase "functional grounded PV system." Also, it should be noted that a functionally grounded PV array using a non-isolated (transformerless) inverter may have the dc array circuit conductors at significant ac voltages with respect to ground when the inverter is operating normally due to the grounded neutral/center tap on the utility 120/240-volt transformer.

System Grounding

"**690.41. PV System Grounding Configurations**

> One of the following system grounding configurations shall be employed:
> (1) 2-wire PV arrays with one functional grounded conductor.
> (2) Bipolar PV arrays according to 690.7(C) with a functional ground reference (center tap)
> (3) PV arrays not isolated from the grounded inverter output circuit
> (4) Ungrounded PV arrays
> (5) Solidly grounded PV arrays as permitted in 690.41(B) Exception
> (6) PV systems that use other methods that accomplish equivalent system protection in accordance with 250.4(A) with equipment listed and identified for the use."

It should be noted that there are no voltage requirements associated with the requirements in 690.41 and that the old 50-volt limit is gone. I suspect it will take some time to figure out what we call these various types of PV arrays. For example, the PV array connected to a non-isolated inverter with a 240 VAC output meets the requirements of (3) because at some point there is a utility transformer that has a grounded 240/120-volt center tap.

Equipment Grounding

Sections 690.43, 690.45, and 690.46 dealing with equipment-grounding conductors have remained essentially unchanged.

Grounding Electrode Systems

This section has again been revised in the 2017 *NEC* and will be discussed in some detail in chapter 6.

PV Circuit Overcurrent Protection and Disconnects

Section 690.9(C) now permits (note: not a mandatory "shall") a single overcurrent protective device (OCPD) to be used in dc PV-source and dc PV-output circuits to protect the modules and circuit conductors even on ungrounded PV arrays where both circuit conductors—positive and negative—are ungrounded. No longer are OCPDs required in both ungrounded conductors. However, the single conductor OCPD in source circuits, dc combiners, and in other locations in the dc system must be in the same polarity conductor. See photo 1.18.

Section 690.13(F)(1) requires that all types of PV systems have a disconnecting means that simultaneously opens all conductors (even those solidly or functionally grounded) of the PV system from all conductors of other systems (photo 1.19). Solidly grounded conductors, in either ac or dc circuits, should not be opened with PV system disconnects. If a solidly grounded dc

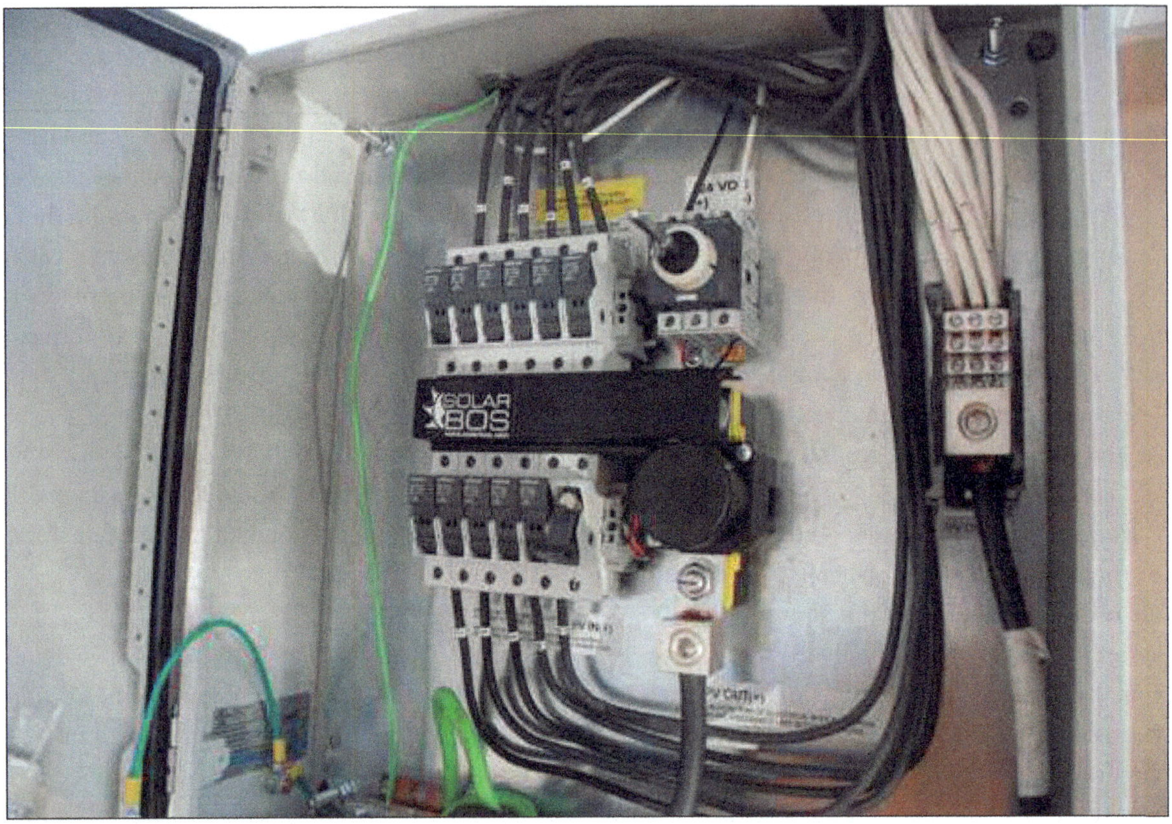

Photo 1.18 • (*top*) DC combiners may look similar for many systems. Although this is a combiner listed under the 2014 *NEC*, fuses are permitted to be in only one conductor. However, to meet 2017 *NEC* 690.15(15) requirements, as an equipment disconnect/isolator, an output switch will be required in both conductors unless one is solidly grounded, and the now functionally grounded conductors should not have white insulation.

Photo 1.19 • (*left*) The DC PV system disconnect or possibly an array isolator may have to be rewired on formerly grounded PV arrays when updating the inverter to a non-isolated unit. A switch pole in each conductor will be required under the 2017 *NEC*.

conductor is opened by a PV system disconnect switch, some circuits could become ungrounded and some conductors marked white may no longer be grounded. If a grounded ac neutral on an inverter output is opened, internal circuits in the inverter (such as filter circuits) may be damaged. This issue will be clarified in the 2020 *NEC*.

Section 690.31(B)(1) allows a white conductor marking on only those circuit conductors that are solidly grounded.

Help for The Systems Designer from The Professional Engineer

On larger PV systems that are 100 kW and larger, some of the general requirements for voltage and current calculations are too conservative for the proper and safe installation of cost-effective PV systems.

Section 690.7(A)(3) allows the professional engineer (PE) to calculate the maximum system voltages based on industry design practices. These might include such factors as microclimates, array mounting devices, and array orientation. This new calculated system voltage may permit more modules to be placed in series than the more conservative 690.7(A)(1) and (2) would allow.

Section 690.8(A)(2) allows the PE to calculate a maximum current that can be no less than 70% of maximum current calculated using the normal method of using 125% of the short-circuit current (I_{sc}). In large systems, these methods using simulated local irradiance may allow smaller, yet still safe, conductor sizes.

Other Changes in Article 690

Scope, 690.1

The scope excludes large PV systems which, are now covered in Article 691.

DC-To-DC Converter Source and Output Circuits, 690.7(B)

This new section establishes requirements for determining the maximum voltage of single or multiple series connected dc-to-dc converters.

PV Rapid Shutdown Systems, 690.12

This section received significant changes for 2017. The previous boundary around the PV array that defined the point where conductors must be controlled to a lower voltage has been decreased from 3 m (10 ft) to 305 mm (1 ft). The length of allowable uncontrolled conductors inside a building has been reduced from 1.5 m (5 ft) to 1 m (3 ft). The time allowance for controlling controlled conductors from initiation has been increased from 10 seconds to 30 seconds. These modified requirements provide additional safety for first responders who must deal with an energized PV array and allow the PV industry, in some cases, easier methods of meeting the requirements. (See photo 1.20.)

There are now three new options that must be used to meet the PV rapid shutdown system (PVRSS) requirements for the conductors inside the PV array controlled boundary [690.12(B)(2)]. These requirements will become effective January 1, 2019.

1. Use a PV array that has been entirely listed or field labeled as a rapid shutdown PV array.
2. Install a PV array with controlled conductors inside the array boundary and not more than 1 m (3 ft) from the point of penetration to 80 V or less within 30 seconds of PVRSS initiation.
3. Install a PV array with no exposed conductors or exposed conductive parts more than 2.5 m (8 ft) from exposed grounded conductive parts or ground.

It should be noted that these three new options will become effective on January 1, 2019. This requirement implies that some sort of module-level control will be required to meet the controlled conductors inside the PV array boundary.

In some cases, module-level power electronics (MLPE), such as microinverters, ac PV modules, or dc-to-dc converters, may meet the PVRSS requirement after they have been listed as rapid shutdown equipment. The 2014 *NEC* required only that the equipment be listed and identified (meaning a listed relay could be used). The 2017 *NEC*, however, requires the equipment to be listed and labeled as a rapid shutdown system or as RS equipment.

The new and revised PVRSS section of UL

Standard 1741 establishes complex and detailed testing requirements. One of these requirements is that the ac output of inverters go to the controlled voltage limit within the required time frame after initiation (usually opening the ac utility disconnect) of the RSS system. This test was not in UL 1741 under the initial RSS requirements.

The initiation device for the PVRSS must plainly indicate the OFF position and the ON position, where the OFF position indicates that the PV array has been put in the controlled state for all circuits controlled by that initiation device. The initiation device shall consist of at least one of the following:

(1) The service disconnecting means.

(2) The PV system disconnecting means.

(3) A readily accessible switch that indicates the ON and OFF positions.

Because various PV systems are installed at different times and may be expanded under different editions of the *Code*, Section 690.56(C) requires a plaque clearly showing a physical plan diagram of the PV system indicating which parts of the PV array are RSS controlled and under which edition of the *Code* that part meets. Section 690.56(C)(1) has detailed marking requirements for the various types of PVRSS.

Finally, the manual reset requirement has been deleted from the 2017 *NEC*.

Arc-Fault Circuit Protection (Direct-Current), 690.11

This section of the *Code* was substantially reduced for 2017. An exception was added that exempts PV systems from the arc-fault circuit protection requirement that are not mounted on buildings. Also exempted are PV systems with PV output and dc-to-dc converter output circuits that are direct buried, installed in metallic raceways, or installed in closed metallic cable trays. This exclusion was necessary because arc-fault equipment at the high current levels found in large PV arrays does not exist.

Photo 1.20 • PV Rapid Shutdown Equipment.
Photo courtesy of SMA

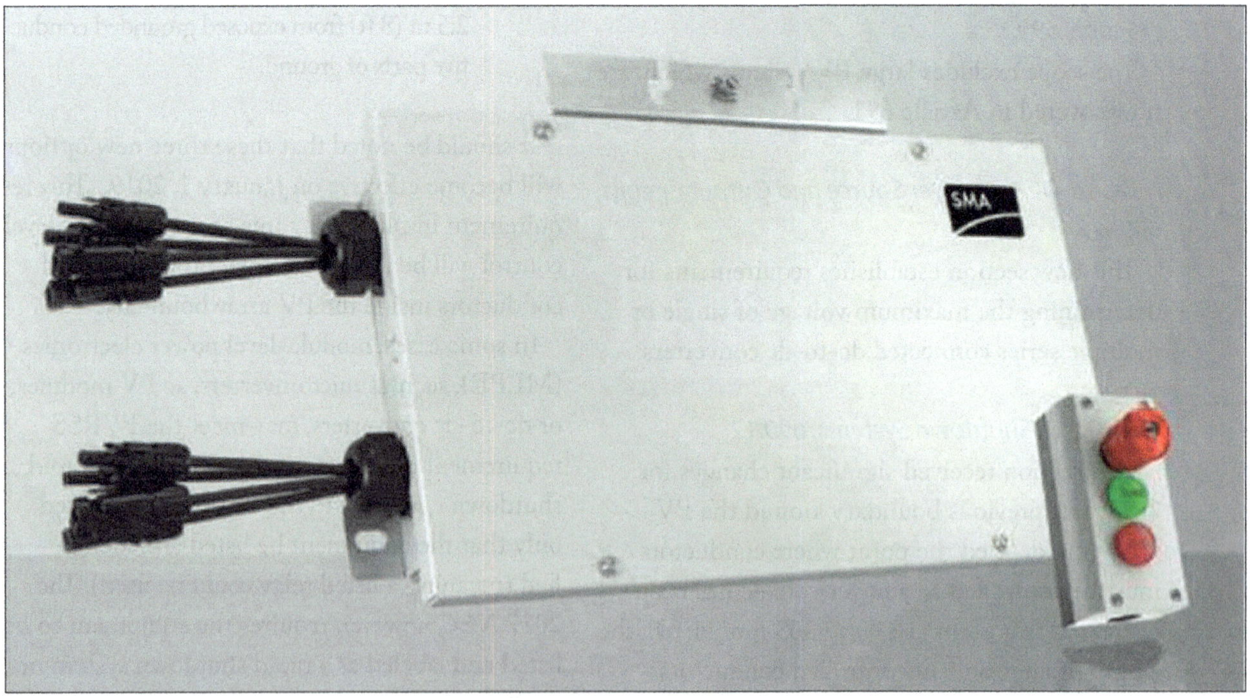

The exclusion does not address larger PV systems on the roofs of commercial buildings. These systems are being addressed using multiple medium-size string inverters with arc-fault detection capabilities mounted in subarrays on the roof.

The manual restart requirement and the requirement for an annunciator have been removed, but are included in the requirements in UL Standard 1741.

Article 691

For large-scale (5 MW or greater) PV systems, not under the ownership or operation of a utility, special design and installation requirements are allowed. These requirements address numerous areas, but in general require a PE-engineered design; limited and controlled access; no local loads except as necessary to operate the generation system; and a full, utility-type substation to make the connection to the local utility. Also, parts of the system that do not comply with the requirements of Article 690 shall be fully justified and documented.

At the request of the AHJ, an independent electrical professional engineer, retained by the system owner, may be required to evaluate the actual installation for compliance with the *Code* requirements and the engineered design.

Even though these large systems (in Articles 690 and 691) are designed by a PE, and may have independent review of the installation for compliance with the design, it is still the responsibility of the AHJ to ensure that the system is safe and meets the various applicable codes. In many cases, it is common for jurisdictions to outsource inspections of large-scale PV systems due to departmental manpower shortages. This outsourcing could be to an independent electrical inspection service or to a licensed professional engineer. Keep in mind that all the large utility-scale systems being installed today use engineered designs.

Scott Humphrey's excellent series of articles on inspecting large PV systems in recent editions of *IAEI News* point out the continuing need for close and detailed inspections of these installations by an AHJ supervising a team of inspectors directly or carefully reviewing the certifications of the independent professional engineer.

Article 705

Definitions, 705.2

Definitions have been added or modified in this section.

A ***microgrid system*** is "a premises wiring system that has generation, energy storage, and load(s), or any combination thereof, that includes the ability to disconnect from and parallel with the primary source." The requirements of this code apply unless the microgrid system is under the exclusive control of a utility.

The term ***utility interactive*** has been modified to just ***interactive*** throughout section 705.

The definition of ***stand-alone system*** has been moved to the new Article 710 – Stand Alone Systems.

Point of Connection, 705.12

Sections (B) and (C) dealing with integrated electrical systems and systems greater than 100 kW have been deleted from this Article. Article 705.12 requirements found in 705.12(A) and the renumbered 705.12(D) [now B)] apply to interconnected electrical power systems of any size.

Point of Connection, Load Side 705.12(B)

This section has been modified to allow other interconnected (interactive) power sources to be connected on the load side of the system disconnecting means.

Center-Fed Panelboards, 705.12(B)(2)(3)(d)

Center-fed panelboards may now be treated like standard panelboards where the backfed PV breaker may be connected at the end (away from the main breaker) on only one of the two busbars

Photo 1.21 • (*top left*) Center-fed panelboard. PV connections now allowed by the 2017 *NEC* at one end only. *Photo by Dan Potkay.*

Photo 1.22 • (*top right*) Residential microgrid system—PV input.

Photo 1.23 • (*bottom right*) Microgrid system—generator input.

and may be rated according to the 120% rule. This rule states that 125% of the inverter output current rating plus the main breaker rating may not exceed 120% of the busbar rating (see photo 1.21).

Wire Harness and Exposed Cable Arc-Fault Protection, 2014 – 705.12(D)(6)

This section dealing with ac arc-fault protection for 240-volt ac inverter output circuits up to 30 amps on flexible, exposed cables has been removed because no suitable equipment is available to meet the requirement.

Disconnecting Means, Sources

Article 705.20 may require some harmonization with the revised system-disconnect requirements in Article 690. These Article 705 disconnect requirements require only ungrounded conductors to be disconnected, whereas the PV system disconnect requirements in 690.13(F)(1) require all conductors to be disconnected. This will be addressed in the 2020 *NEC*, where only ungrounded conductors shall have disconnect requirements.

The equipment disconnecting requirements in 705.21 are consistent with the equipment disconnecting/isolating requirements in 690.15 because both require only the ungrounded conductors to be disconnected.

Disconnect Device, 705.22

This slightly revised section allows power-operated disconnects to be used, but this allowance for PV system disconnects in the 2014 *NEC* was removed when 690.17 was deleted. Again, harmonization may be required in future editions of the *Code*.

Photo 1.24 • Microgrid research project with utility-interactive PV (array on roof), energy storage, and smart receptacle outlets measuring energy and controlling smart appliances.

Microgrid Systems, 705 Part IV

This part has four sections that outline the requirements for operation of a microgrid and connection of that microgrid to a primary power source. These requirements are based primarily on the operation of utility-interactive inverters found elsewhere in Article 705 and previously in Article 690 (photos 1.22, 1.23, and 1.24).

Wiring Systems, 705.175

This section has been moved to Article 710, Stand-Alone Systems.

Article 706, Energy Storage Systems

This new article was added to the 2017 *NEC* and contains nearly all the references to batteries or other energy storage systems previously found in Article 690. Article 706 is lengthy and has requirements that were not previously in the *Code*, such as a requirement to determine and mark the rated short-circuit current of the energy storage system. Article 480, Storage Batteries, remains in the *Code* and has some of the same new or revised requirements that appear in Article 706. Chapter 5 will cover Article 706 in detail.

Article 710, Stand-Alone Systems

This new article gathers requirements previously located in sections of Articles 690 and 705 relating to stand-alone systems. The Article primarily addresses the requirements for a remote, off-grid dwelling or other load and there is considerable flexibility allowed in the characteristics of the power source supplying that load.

Article 712, Direct Current Microgrids

Direct current distribution systems may be fed by a PV system or another dc source. The grounding terms defined in this Article differ somewhat from the new grounding terms found in Article 690. Section 712.2 defines a ***grounded two-wire dc system*** and a ***grounded three-wire dc system*** as having a solid or reference ground between one circuit conductor and the equipment grounding system. It also defines a ***reference-grounded dc system*** as one that is not solidly grounded, but has a low-resistance electrical connection to ground that maintains the circuit conductor at equipment grounding system voltage under normal operation. This sounds like the Section 690.2 definition of ***functional ground***, but could use a circuit breaker or fuse as the low-resistance element. Section 712.2 also defines ***resistively grounded*** as a system having a high-resistance connection between the circuit conductor and the equipment grounding system. Of course, ***ungrounded*** is defined and it (to a certain extent) agrees with that term in Section 690.2.

On a dc PV system, we should certainly be well "grounded" between these two Articles—I think. (See photo 1.25.)

UL Safety Standards

UL Safety Standards are developed to ensure that equipment certified (listed) to these standards can be installed according to the installation requirements found in the *NEC*, thereby yielding an essentially hazard-free electrical

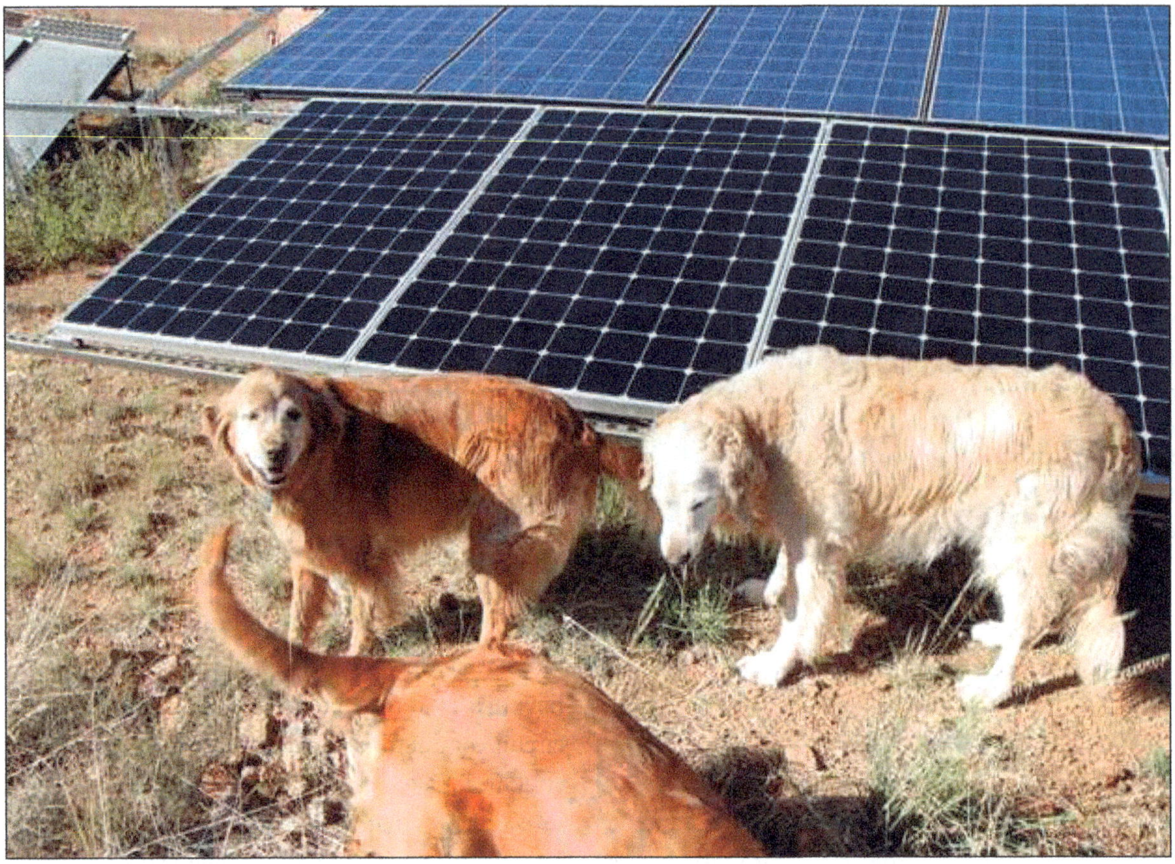

Photo 1.25 • "We just know there is a 'function ground' here somewhere. Give us time; we'll find it."

installation. In the days preceding electronic power equipment (yes, that was even before my time), application of the *Code* requirements was relatively straightforward: connect conductors to passive devices like switches, overcurrent devices, and transformers. Many electricians could do that with a little training, and even the inspections were relatively easy.

However, AHJs are now faced with active electronic power-processing equipment scattered throughout our electrical power systems, and PV systems are no exception. AHJs must now ensure that active inverters, active charge controllers, rapid shutdown equipment, dc PV arc-fault circuit interrupters, microinverters, and ac PV modules plus module-level electronics are properly installed and that they will operate safely.

Many of these devices have multiple inputs and outputs and require not only electrical connections to the rest of the electrical circuit, but also careful adjustments of software- and hardware-driven internal parameters to operate safely.

Each of these active devices, made by numerous manufacturers, has different requirements for safe installation. It is neither possible nor practical to put detailed instructions in the *Code* for the installation and adjustment of each of these types of devices. The *Code* would become far too ponderous to deal with in its present form.

For this complex equipment, the best the *Code* can do is to include Section 110.3(B), which requires that the instructions and labels provided on certified (listed) equipment be used and followed to properly install that equipment.

This requirement in the *NEC* places a significant burden for ensuring the safe installation directly on the quality and detail of the equipment installation instructions that are required by the UL Standard for that equipment. These standards must be detailed and understandable because several Nationally Recognized Testing Laboratories (NRTLs) other than UL such as TUV, ETL and CSA will be using them to test and evaluate PV equipment. (See photo 1.26).

These requirements lead us to the effort required to develop these standards. Although Underwriters Laboratories publishes the Standards, the various UL Standards Technical Panels (STPs) write them. The STPs are composed of balanced volunteer groups of manufacturers, AHJs, technical experts, users, government agencies, and general interest individuals. In some cases, these groups may include more than fifty people. There are standards and STPs for each of the major categories of equipment in PV systems. These include UL 1741 for inverters (including charge controllers, microinverters, rapid shutdown equipment, and ac PV modules); UL 1703 for PV modules; UL 1699B for dc PV arc-fault circuit interrupters; UL 4703 for single conductor PV cable; UL 6703 for PV connectors; UL 2703 for PV racks; UL 489B for dc PV overcurrent devices; and others.

In many cases, the requirements in the Standards are so complex that the entire STP does not, as a group, develop them. Various working groups are set up composed of STP members and other technically competent interested parties to draft sections of the Standard. After those drafts are edited to comply with UL writing standards, they are presented to the STPs for review, final refinement, balloting, and integration into the whole Standard. An example is the draft Standard for the *Code* requirement in 690.12 for a PV rapid shutdown system (PVRSS).

Underwriters Laboratories' engineers developed the basic outline and the initial contents for the draft PVRSS Standard and then published them as a Certifications Requirements Decision (CRD) in March 2015. UL then formed a working group outside of the STP to further refine this draft. The working group, consisting of more than fifty people, met on a weekly basis via webinar for more than twelve months to investigate, evaluate, and develop the require-

Photo 1.26 • Rigorous application of standard requirements will minimize issues like this failed PV module solder bond.

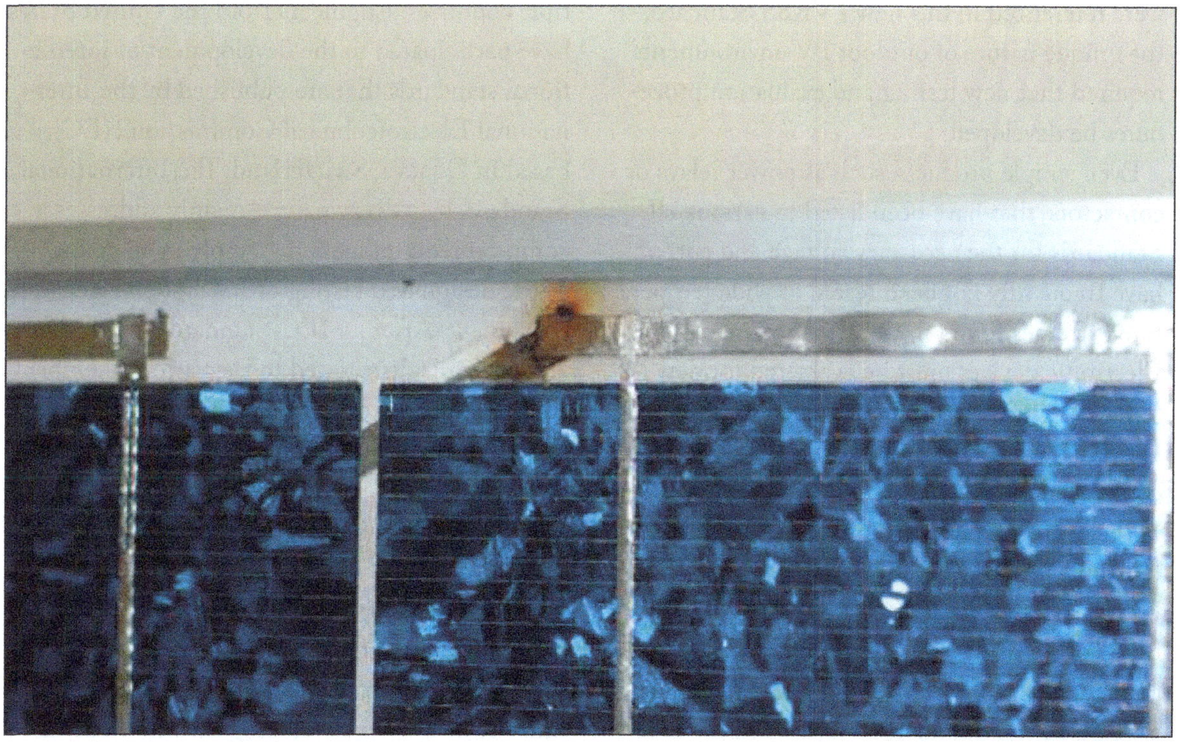

ments for the PVRSS based on the very limited requirement stated in Section 690.12 of the 2014 *NEC*. And as the PVRSS requirements in the 2017 *NEC* became known, the draft Standard was again revised.

The working group had to consider all possible ways in which various equipment manufacturers could implement the *Code* requirement. Because this PVRSS system provides a life safety function, the testing and evaluation criteria that the Standard required would have to be rigorous, extensive, detailed, and very complex. The PVRSS system may have to work properly in an outdoor PV system environment that includes extremes of temperature, humidity, UV irradiation, wind, snow, ice, and blowing dirt. The standard must require testing to evaluate PVRSSs and PV rapid shutdown equipment (PVRSE) under these conditions plus surges that might occur from nearby lightning strikes on the connected ac and dc circuits. The PVRSS equipment must also operate reliably for the 40- to 50-year life of the system. While there are UL Standards to evaluate some of these conditions, and those standards were referenced in this new PVRSS Standard, the unique nature of outdoor PV environments required that new testing and evaluation procedures be developed.

Even simple products, such as power relays or contactors, that have been listed to various UL Standards for the necessary voltage and current have frequently not been tested in wide temperature extremes found in the PVRSS application. Electronic devices must be evaluated for new types of failure modes such as operation in high humidity environments and the response when control circuits fail or are interrupted.

The draft Standard is more than 20 pages long and, after review and modification by the UL 1741 STP, it will be added to the UL 1741 Standard (already 144 pages long).

In addition to the changes made to include PVRSS in UL Standard 1741, an entirely new standard was developed to address systems that control hazards within the boundary of the PV array. This Standard, UL Standard for Safety for Photovoltaic Hazard Control (UL 3741), will be published in late 2018 or early 2019.

Harmonization with International Standards

Photovoltaic equipment is frequently designed and manufactured to be sold and installed in international markets. The equipment must meet the codes and standards of each country. While there is some variation in the electrical codes between countries that have such codes, the connections of PV equipment to electrical systems is not too different from country to country. Some editions of the *NEC* have been translated into other languages (photo 1.27). Unfortunately, the standards governing how equipment is built and tested can differ significantly between countries. There has been a long-term, on-going effort to harmonize safety and construction standards of PV equipment between countries.

A harmonized standard would reduce manufacturers' costs of certification and listing in multiple countries. Engineers from the United States have participated in the development of international standards that are published by the International Electrotechnical Commission (IEC), based in Geneva, Switzerland. The International Standard, by its very nature, mainly addresses common requirements that apply in all countries to the design and testing of PV system electrical equipment. After an IEC Standard is published formally, it is distributed to the various member countries where they evaluate it and add sections to the standard that elaborate on the "Country Differences" that will be used along with the common International Standard requirements to evaluate products in a particular country.

As United States safety standards become harmonized with the international standards, working groups in the United States spend two to three times as long developing the international

Photo 1.27 • Codes and standards are being harmonized between countries

standards as they would a U.S.-only standard. The working groups first develop the United States standard, and it is published as a UL Standard. They then work on the international standard and, finally, they develop the "Country Differences" when the international standard is published. This is a very long process involving possibly hundreds of people and thousands of hours. All of this is to ensure that PV equipment can be installed according to the requirements in the *NEC* and will operate safely when so installed.

The AHJ should now have a better idea why the instructions provided with a listed product are critical and why those instructions must be reviewed during the permitting, plan review, and inspection processes to ensure safe and *Code*-compliant installations of PV systems.

Technology Advances

Electric Vehicles

There are now both plug-in hybrid electric vehicles (PHEVs), having fueled engines plus electric motors and batteries, and pure electric cars (EVs), having electric motors and batteries, that require charging stations at the home base of such vehicles and in locations throughout the area where they'll roam (photo 1.28). Similar to cell phone coverage, charging stations will be concentrated in metropolitan areas and then will spread to less populated areas as the demand for extended coverage grows. Owners of electric vehicles will certainly have charging stations in their homes and probably at their job sites. At the very least, there will be a new type of receptacle outlet to manage and probably relatively high-current branch circuits. Article 750, Energy Management Systems, has been expanded in the 2017 *NEC*. In addition, a new Article 706 applies to charging systems for electric vehicles and energy storage systems for PV systems.

Plans are also being made to have parked, fully charged electric vehicles feed some of the energy stored in the onboard battery bank back into the utility grid at peak demand times. Controlling this exchange of energy from grid to vehicle and back, and ensuring that the car is ready and charged when needed, will require communication between the car, the owner, and the utility. Such communication links may be wireless, over the Internet, or through a hardwired connection along with the power connections. Like utility-interactive PV systems, these vehicle storage systems will require new *Code* changes and additional inspections to ensure public safety.

Energy Storage Systems — Large and Small

Utilities will embrace the dispatchable energy storage and generation systems. They will be able to tap energy that has been stored or that is available throughout the distribution network for use to offset peak demand loads. This operation will avoid having to increase the size of already taxed power plants and transmission lines (photo 1.29). Backup generators at hospitals and other locations are already being used in this mode of operation. These emergency power systems are leased, operated, and maintained by third parties

Photo 1.28 • (*top*) Electric vehicles are back after a century of disuse in various forms, including EVs, hybrids, and fuel-cell vehicles. They will dominate the transportation industry in the future.

Photos 1.29a and b • *(middle and bottom left)* The infrastructure of utility generation and distribution systems is fairly robust, but very old, and somewhat inflexible in dealing with increased use of energy sources and its associated issues. Smart Grid programs are designed to modernize the entire system from the generation plant to end-use load.
Photos by John Watson.

who run them when not needed for emergencies and sell the power to the utilities during peak load periods. New Article 706, Energy Storage Systems, addresses these systems, and most of the battery requirements in Article 690 have been moved to this location.

Flow batteries are coming to the market. These batteries use stored liquid chemicals and a process that yields a long-lived battery that can be rapidly charged and deeply discharged virtually an unlimited number of times. The batteries will be charged, and energy will be stored, during off peak demand periods and released back to the grid during peak demand times. Of course, the process will require utility-interactive systems to interface with the utility grid and communications systems to control the process. These systems, wind turbines (*NEC* Article 694) (photo 1.30), and fuel cells (*NEC* Article 692) operating from natural gas will probably first appear in commercial buildings that have the necessary space. These systems will either be leased or owned, but in most cases these new technology systems will require permitting and inspections of the added mechanical systems, the utility-interactive electrical connections, and the communication circuits.

Photo 1.30 • Wind turbine farms are providing significant amounts of alternative energy today. *Photo courtesy US DOE.*

The Smart Grid

Energy demands throughout the country, and the world, are increasing steadily and will necessitate some combination of increasing the supply from new generation plants (coal, gas, oil, or nuclear and renewable), reducing the demand through conservation, and restructuring of the existing distribution and consumption system. Climate change considerations and market conditions will have significant impact on which energy sources will predominate in the mix. The infrastructure of utility generation and distribution systems is fairly robust, but very old, and is somewhat inflexible in dealing with increased use of distributed energy sources and issues associated with moving power from the sources to consumers in other areas. The Smart Grid programs are designed to modernize the entire system from generation plant to end-use load.

Although many see the term *Smart Grid* and think that it will not impact the premises wiring, the *NEC*, or the inspection process, that would be a misconception. Utilities are now installing smart meters as rapidly as they can find funds to do so. These smart meters are computer (microprocessor)-based and not only allow remote reading and power quality recording (real power, reactive power, power factor, and more), but may also serve as the interface between the Smart Grid and the smart house. Some smart meters can even allow power to be remotely disconnected when bills are not paid.

Smart houses will soon become a reality. Appliance manufacturers are already making dishwashers, clothes washers, and other appliances that communicate through either hardwired or wireless communication systems to the homeowner, to the smart meter, and (in some cases) to the utility. When financially beneficial to the consumer, or possibly when legislated, these smart appliances will be remotely controlled by the utility so they may be operated only during times of low demand on the utility system. These appliances may have unique plugs and receptacles and possibly communication connections. All of these new load connections must be inspected, of course.

What will be the impact on the *Code* of a house that has load circuits and loads that may be remotely controlled or managed? How will service, feeder, and branch circuit sizes be determined? Copper conductor prices may rise so high that the electrical equipment industry is forced to control power flow so that smaller conductors can be used. Eventually, the use of electricity on the premises may be scheduled so the maximum current ever drawn may be significantly less than that requiring a 100- or 200-amp service today. Smaller conductors and circuit sizes may reduce the ever-increasing costs of electrical installations, but *Code* revisions would be needed. With the demise of the incandescent light bulb, do we really need three volt-amps per square foot for general-purpose circuits? Oh yes, there will be those 100+ inch flat panel displays on all four walls to deal with.

Is dc coming back?

There is a trend of going back to dc end-use appliances. Most electronic appliances such as cell phone chargers, radios, TVs, DVRs, DVD players, cable boxes, satellite receivers, track lighting and the like—while being plugged into a 120-volt ac receptacle outlet—run on low-voltage dc. Fluorescent and LED lighting bulbs and fixtures also operate on direct current. Significant losses are incurred to transforming the 120-volt ac line voltage into low-voltage dc. New Article 712, Direct Current Microgrids, describes systems and circuits that may or may not be directly connected to the utility grid. There are also color codes for dc branch circuits in Article 210.

At the present time, dc lighting fixtures are being installed in commercial buildings and are powered during the day directly from PV power systems with no conversion to ac until the electronic ballasts are reached. Solar lighting power is supplemented with utility power when necessary.

With the reduction in use of incandescent light bulbs over the next few years, the return of low-voltage dc power distribution systems for lighting and electronics is almost a certainty. Shades of the 1970s and 1980s! Maybe those off-grid, long-haired solar hippies who insisted on staying with the 12-volt dc PV systems and electrical systems in their homes were far ahead of their time! Of course, appliances requiring significant power for heat or mechanical motion like ranges, clothes washers, toasters, water heaters and the like will usually need higher voltages to keep the current and hence the conductor sizes to reasonable sizes. But then there are heat pump water heaters, induction ranges, and ultrasonic washers that operate more efficiently than conventional appliances.

Renewable Energy Systems

Large wind-power systems have been installed for many years, and many of those systems are not owned and operated by utilities on utility properties. Therefore, they come under the requirements of the *NEC* and should be inspected for safety using the requirements of Article 694, Wind Electric Systems. UL also has standards for large and small wind turbines. Photovoltaic power systems for residential and commercial use have been around since the mid-1970s with substantial growth starting in the late 1990s.

While ever-increasing numbers of residential and small PV systems are being installed throughout the country, real power production will come from the numerous megawatt commercial systems being installed and planned. Systems as large as 2000 megawatts are being planned and installed. Some of these will be solar thermal systems along with the PV systems. In many cases, these large systems are said to be "Behind the Fence" and not subject to the requirements of the *NEC* and inspections, but they are mainly owned and operated by private companies under power purchase agreements (PPAs) and should be fully *NEC* compliant. New Article 691, Large-Scale Photovoltaic (PV) Electric Power Production Facility, addresses these larger (over 5 megawatts) systems.

True ac PV modules with microinverters bonded to the back of the PV module with no dc wiring subject to *Code* requirements. They are on the market in catalogs and in big box stores at impulse-buying prices and are being installed by PV professionals. Will these products be listed? Will these be permitted? Will they be installed by qualified persons (those having the skills, training, and experience as per Article 100)? Will they be installed on dedicated circuits that will ensure public safety? Or will they be plugged into the nearest GFCI outlet and violate *Code* regulations and safety in many ways?

Of course, mainstream manufacturers are making and selling ac PV modules through normal solar distributor networks and they are properly listed to UL Standard 1703 (module) and UL Standard 1741 (microinverter). They are being installed by professional, licensed PV installers and other qualified individuals and organizations.

Dc-to-dc converters attached to or connected to PV modules are well established in the market along with (in some cases) matching inverters. Due to the nature of their operation, many of the *Code* dc requirements do not apply to these devices. The instruction manuals for these listed products must be used to determine compliance [110.3(B)]. The *NEC* does not specifically give guidance on how to deal with them, and future editions of the *Code* may show a similar trend.

All of these changing and emerging technologies will create challenges for inspectors and plan reviewers and also opportunities to excel.

The Challenges

Electrical inspectors, plan reviewers, and combination inspectors are being challenged today (and will be for the foreseeable future) with these evolving energy production and storage sources that will be in use throughout the country. Many of these systems will appear connected to premises wiring and they will come under the requirements of the *NEC*. Many multi-megawatt PV, wind, and solar electric farms will fall under the *Code*.

The Environment

With a salute to the U.S. Postal Service, neither snow nor rain nor heat nor gloom of night stays these PV systems from producing power. Well, OK, the PV system will stop producing power at night and will have reduced output when the sun is not shining brightly, but when the sun next comes up, it will be ready to produce the rated power (or more) day after day and year after year.

Environments in which PV arrays operate are harsh. The array, module, and string wiring are exposed to environmental conditions from the coldest temperatures in the winter to the hottest temperatures in the summer—plus the added stress of high winds, sleet, hail, snow, and, nearly every day, additional heating due to solar radiation.

A PV system may be subjected to high levels of sunshine (irradiance) on a continuous basis (for three hours or more) when the sky is clear, the sun is bright, and there is low atmospheric pollution (humidity, dust, smog, etc.). Under those conditions, a PV array may produce power and current that can reach 125% of the rated output. A PV system is designed to accommodate these situations.

A PV system will also be subjected to short periods (minutes) of cloud-enhanced, snow-enhanced, or water-enhanced irradiance that can be up to 150% of that on a bright, sunny day.

These environmental factors, plus the fact that this is an electrical power system that produces power in a daily cycle, means that all the conductors, any combiners, all overcurrent devices and disconnects, and each and every electrical terminal are subjected to a cyclic flow of current from zero at night to a maximum at or above (for short periods) the rated value of the device every day. Additionally, environmental heating and cooling variations on daily and seasonal cycles impose additional stresses on PV systems.

This high-stress environment and operation

Photo 1.31 • Close-mounted PV arrays allow little room for inspecting the module and array wiring.

is unique compared with the typical residential and small commercial electrical systems that are familiar to the AHJ. When considering that these systems may operate unattended and unmaintained for years—if not decades—PV systems demand the highest quality design, material, equipment, and installation. In the plan review and final inspection of a PV system, the AHJ has the final responsibility and authority to verify that the stringent requirements of the electrical and building codes are followed in the installation. There are some details in various PV subsystems that may need additional attention to achieve a hazard-free, long-lived system.

Yes, Time and Money Are in Short Supply

It is acknowledged that many residential electrical systems are inspected in less than an hour. This is probably acceptable for an electrical system that has remained essentially unchanged for 100 years; where the inspector is quite familiar with *Code* requirements for those electrical systems; and where the inspector is also familiar with the technologies and equipment being installed. Experience with the contractors involved also speeds the inspection process.

It is also acknowledged that many jurisdictions do not do plan reviews and, unfortunately, this hampers the ability of the inspector to easily verify all *Code* requirements. Photovoltaic modules on roofs are generally mounted close to the roof and it is generally not possible to examine all the connections in the PV array, including conductor size, grounding and bonding, and module wiring interconnections. (See photo 1.31.) A related issue in a well-installed PV or other electrical system is that it is sometimes difficult to verify conductor types and conductor sizes where connections are made inside standard disconnecting devices and inverters. These particular *Code* requirements are best reviewed in a plan review stage even if that review is only a half hour in the office looking at the detailed schematics and diagrams that were submitted with the permit application. If issues are found during the office review, it is almost certain that similar issues (or more critical ones) will be found if time is taken to examine those areas during the field inspection.

While many larger PV systems are ground mounted, the great majority of residential PV systems, and smaller commercial PV systems, are mounted on the roofs of buildings. Unfortunately, some jurisdictions have insurance policies that prohibit inspectors from going on the roof, making examinations of those critical PV system components difficult, if not impossible. If a ladder is available, and the installing contractor makes all areas of the array easily accessible by providing ladders, the inspector might be able to view some of the array from the ladder without going onto the roof. The roof is no place to be in inclement weather.

The Array-Mounting and Mechanical Considerations

Combination inspectors who are also building inspectors are probably more aware of the mechanical requirements to fasten PV modules to roofs the building structures than typical electrical inspectors might be. Most mounting systems, including PV racking systems, require that the PV array be connected to the structural elements of the roof and not just to the thin roof sheathing. Without going into the attic (and the flat-roofed homes common in the southwest United States that have no attic), it is difficult to verify that mechanical fasteners are indeed being inserted into the structural roofing components (photo 1.32). In many parts of the United States, standard spacing for trusses is 24 inches with older homes having structural members at 16 inches. A quick check of the exterior mounting feet for the array rack should show them to be consistently on some multiple of these dimensions. Racking and mounting systems are continually evolving, and any system must be installed according to the manufacturer's instructions.

The module interconnect wiring should be firmly secured to the racking members or otherwise supported. Those exposed, single-conductor cables should not be hanging loosely below the modules where they are subjected to additional mechanical stress from wind loading and the possibility of abrading the insulation against the roof. Unfortunately, many of the plastic wire ties that are frequently used to attach these conductors to racking members, although listed for outdoor exposure, do not have the necessary longevity for the years of service needed in a PV

Photo 1.32a and b • Roof trusses at 24 inches on center (more or less). But where are they after the sheathing and shingles are installed?

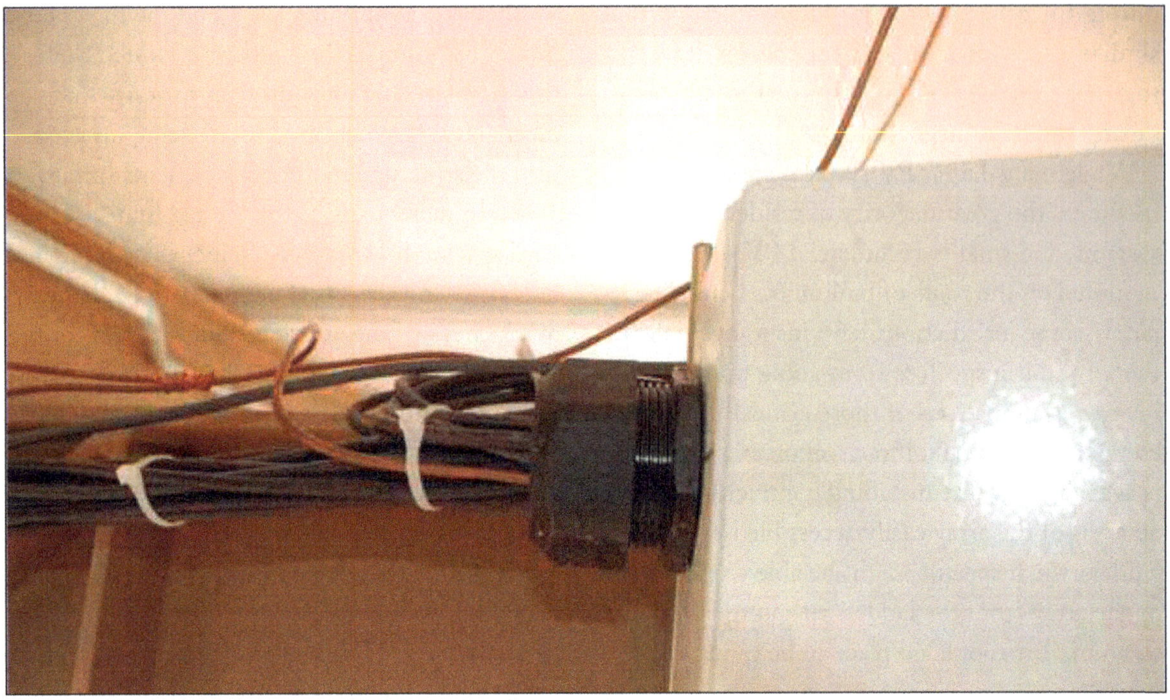

Photo 1.33 • Plastic cable ties are not durable in the PV environment, particularly these white ones that are not rated for outdoor use.

system. (See photo 1.33.) It is suggested that metal clips with rounded edges be used to attach these wires to provide the longevity needed for a PV system.

Module connectors should be firmly seated and, to some extent, able to be visually evaluated at a distance. But where the conductors have been properly secured to the racks, the connectors may be less visible.

The grounding and bonding methods used for the modules and the racks should be evaluated and compared with the instructions for those devices.

Can't Get on The Roof?

In jurisdictions where it is not possible for the AHJ to get to the roof to closely examine the aforementioned parts of the array closely, it is possible to query the installer about some of the things mentioned above. For example, the method of grounding and bonding the modules and the racks can be discussed while looking at the instructions for those pieces of equipment. The ties or clips used to attach and secure the module wiring to the racking members could be discussed, along with any listing or other information that addresses the durability of those devices in the PV application. The installer might be queried on the method of determining where the racking- or module-attachment devices were located in relation to the structural members of the roof.

Torqueing Considerations

All electrical connections that require a threaded fastening have a torque requirement for making that connection correctly. Threaded screw or bolt connections are found on disconnect switches, overcurrent devices, dc combiners, and inverter inputs and outputs. Because of the daily cycling of current in these circuits, it is critical that they be torqued to the manufacturers' specifications. The installer should be able to provide data on all torque specifications. The torque values that should have been used should be readily available in the manuals and cut sheets provided with the permit application.

The installer should be able to show the AHJ

Photo 1.34 • All of this equipment has terminals, terminals, and more terminals. Each terminal must be properly torqued.

Photo 1.35 • Do not attempt to make a supply-side connection inside the meter main combo. The listing on the load center would be violated.

the torque wrenches and torque screwdrivers used to make these connections. A visual inspection should be made to see if all conductors are properly inserted into terminals and the AHJ should have the PV installer demonstrate on a few randomly selected terminals that they have been properly torqued [*NEC* 110.14(D), Informative Annex I, *NECH* Exhibits 110.3, 110.4]. Also, the torque tools can be used to verify that the minimum value of torque has been achieved. Obviously, no torque test should be done on energized circuits.

The AC Utility Connection

Although Section 705.12 of the *NEC* is being improved with each edition of the *Code*, the requirements still confuse some installers and AHJs. Following are a few simple checks that can be done in the field to at least partially verify *Code* compliance.

A supply-side connection, as allowed by 705.12(A), will be made between the existing revenue meter and the main disconnect on the service. This connection is typically not allowed to be inside a meter main load center combination where tapping either the internal conductors or the bus bar between the meter socket and the main disconnect would violate the listing on the load center (photo 1.35).

Photo 1.36 • PV supply-side meter adapter.
Courtesy of San Diego Gas and Electric.

There are now several listed meter socket adapters on the market designed for making a supply-side connection at the meter. The existing meter is removed, the adapter is installed in the existing meter base socket, and the meter is reinstalled by plugging it into the adapter. The adapter provides a set of terminals that are a supply-side connection. This adapter might conceivably even be used in a meter main load center combination to make a utility connection without violating the listing on the load center. While this is a listed device, utility approval is usually required for its use because many utilities consider the meter and the meter base to be under utility control. (See photo 1.36.)

The rating of a supply-side connection for a PV system is limited by the size of the service entrance or any meter adapter or conductors involved with the service [705.12(A)]. A 100-amp service entrance that has 100-amp meter and 100-amp service-entrance conductors could be fed by a 100-amp PV overcurrent device. If a meter adapter were used, it would also have to be rated at 100 amps to use the full 100 amps from a PV system.

Load-side connections, under 705.12(B), are made by backfeeding a circuit breaker in the main load center or making a connection elsewhere in the electrical system at a subpanel or possibly even by tapping a feeder. Although 705.12(B) has been clarified in the 2017 *NEC*, there are still a number of considerations to be made that are best checked in the office during a plan review exercise. However, the most common connection is a backfed breaker in the main load center and several items are relatively easy to verify for *Code* compliance. (See chapter 7 for the details on utility connections.)

Continuing Education and Information

The challenge for every electrical inspector and plan reviewer is to keep abreast of new developments as they start to appear in residential, commercial, and industrial electrical systems. Inspectors and the plan reviewers need to know as much, or more, about these devices and systems as the people installing them. That has been true in the past and it needs to be the standard of performance in the future if the inspection community is to ensure the safety of the public.

Where the *Code* and the standards cannot keep up with these new systems and devices, inspectors and plan reviewers must devote time to educate themselves on the systems that they are and will be inspecting. Strong continuing education programs for inspectors and plan reviewers must be a part of the planning in every jurisdiction. Time and funding should be budgeted for classes, webinars, technical documents, and the equipment needed to efficiently and proficiently review and inspect these ever-changing electrical power systems.

Inspectors and plan reviewers should have electronic copies of all codes, handbooks for those codes, and technical data (including manuals and specification sheets) for all types of systems being inspected (and for equipment that may be installed on those systems). Laptop computers, with screens that can be read outdoors, with this information (updated as necessary) should accompany each inspector as field inspections are conducted. Communication between inspectors and plan reviewers on a real-time basis via cell phone and wireless computer link will be required. Using a digital camera and downloading and transmiting on-site pictures will be necessary.

Facts of Life

More Dangerous. Photovoltaic systems are potentially more dangerous than typical residential or commercial electrical power systems. With voltages up to 600 volts dc on residential installations and up to a 1000 volts dc on commercial installations, the voltages are higher than those of typical electrical systems. If energy storage systems are involved, currents may also be higher than those found on non-PV systems. Energy storage systems in factory-sealed containers may be operating at voltages up to 600 volts dc in residences.

More Complexity. Photovoltaic systems are more complex than typical residential or commercial electrical power systems and that complexity is continually increasing due to technology changes and changing *Code* requirements. There are now frequently changing requirements for PV rapid shutdown systems to protect first responders (690.12). We have dc PV arc-fault circuit interrupter requirements in 690.11. There are also PV utility-interactive inverters with changing requirements and designs due to new utility requirements in some states to provide PV "ride through" of utility disturbances, including low voltage (down to 50% of nominal) and frequency variations.

Installer Qualifications. While many PV systems are being installed by competent individuals and organizations, there are still numerous systems being installed by people who have less than full competency in the required knowledge base and skills. Homeowners can install PV systems in some jurisdictions with no permits or inspections other than what the utility might potentially require. Unqualified electricians, who have had no specific PV training, are installing PV systems. New "PV installation companies" are popping up left and right to take advantage of the boom in PV installations without being fully trained to do those installations properly. As an aside, as I write this, I received a call from an electrician who had been hired to install a PV system for a residential customer who had bought the equipment himself. The electrician had absolutely no knowledge of how to do the job or that there were any requirements in the-*NEC* for such installations. It also appears that the customer may have purchased an unlisted inverter on the Internet where all things are available.

AHJ Qualifications. We don't like to admit it, but just as with some PV istallers, some authorities having jurisdiction and local inspection requirements are not always up to the task of ensuring that installed PV systems are safe and meet the latest *Code* requirements and safety standards. Proper training and education for local AHJs on the latest technology in PV systems is critical for the safe operation of those systems for many years to come.

The Electrical Specialist. There are AHJs in the inspection community who have come up through the ranks as master electricians and have additional training in PV systems. They have been working with PV systems daily since the beginning—some since as early as the mid-1980s. These people are the electrical specialist inspectors. These individuals are frequently involved in PV more than just at the inspection level. They may be members of code-making panels for the *NEC*; members of UL Standards Technical Panels for PV equipment; active members in local IAEI chapters and IAEI sections; and may even be writing articles in *IAEI News*.

The Combination Inspector. At the other end of the spectrum, the latest trend in the inspection community is to hire a "combination" inspector who will be responsible for enforcing several disciplines, such as plumbing, electrical, mechanical, and building codes. The theory is to hire one inspector instead of two or more due to budget limitations and emphasis on other local issues. This combination inspector may have come from the plumbing industry and is now inspecting not only plumbing systems, but also electrical power systems, including PV systems.

This individual will typically not have a firm background in the *NEC*, let alone the very specific safety requirements associated with PV systems. Combination inspectors should expect and have been provided with training in the basic requirements of the *NEC* (its first four chapters) and in basic electrical theory—and where at all possible be mentored by a senior electrical inspector through on-the-job training.

Individuals who find themselves lacking PV-system training or mentorship are typically not to be blamed. They are merely fulfilling an obligation placed on them by their employers. Consideration should be given by the jurisdiction to hiring additional electrical specialist inspectors or they could contract with professionals who specialize in PV systems where possible.

The Permitting and Inspection Process. The process for permitting and inspecting PV systems varies substantially throughout the United States. In some areas of the country, no inspections are conducted on PV systems. In other areas, a permit is nothing more than a form filled out noting the location of the electrical system and a paid fee. In some states, there are no inspections of residential PV systems, and the utility is responsible for ensuring the quality of installations before allowing the interconnection.

On the other hand, many jurisdictions wisely require a full permitting package (with diagrams), a plan-review stage, and a detailed inspection before approving the system as *NEC* compliant. The utility then performs their required inspections prior to interconnection. After the system has been powered up, there may be a final inspection by the AHJ.

An expedited permitting process for PV systems less than 10 kW was developed by the Solar America Board of Codes and Standards (now defunct) and a copy can be downloaded from the SolarABCS web site: *http://solarabcs.org/about/publications/reports/expedited-permit/index.html.*

Numerous other informative documents relating to PV systems are available on the SolarABCs website at *http://solarabcs.org/index.html*

Chapters 8, 9, and 10 cover the plan review and inspection of PV system in greater detail.

Budget and Administrative Restrictions. In many areas of the United States, funding is limited for inspection organizations. Combination inspectors are being hired more frequently than specialist electrical inspectors. Budgetary and manpower restrictions are forcing inspectors to do more inspections per day and, in some cases, these restrictions allow only 15 minutes for an inspection. A thorough inspection of a PV system might require two or more hours, including plan review. In many areas, this amount of time is just not available to the inspector.

Training is also an issue for AHJs who inspect PV systems. Funding is frequently not available, nor is time allocated to properly train the inspectors who will be working with the constantly changing requirements in the *Code* for PV systems. To ensure the safety of the public, inspectors should at least be as familiar with PV systems as PV installers and PV system designers. This familiarity requires periodic and continuing training for inspectors.

It should be noted that PV systems and PV equipment are becoming complex to the extent that instructions for installing these systems can no longer be fully included in the *NEC*. The instructions for installation of that equipment will be found only in the manuals for the listed equipment. These instructions can be, to say the least, quite complex; far more complex than can be put in the *NEC* [110.3(B)].

Moreover, for unknown reasons, some jurisdictions have insurance policies that prohibit AHJs from going onto rooftops to inspect not only PV systems but also HVAC equipment. This prohibition is particularly troublesome where many commercial PV systems are roof-mounted and 90% of the PV equipment that should be inspected is on the roof. In the southwestern United States, with common flat-roofed and low-sloped-roof buildings, PV systems (and sometimes inverters) are frequently mounted on rooftops. Of course, with ac PV module system or microinverters, nearly all systems that should be inspected are found on rooftops.

Resources

There are many resources available to inspector and plan reviewers. Most equipment manufacturers have electronic downloadable PDF files of manuals that will be useful. Here are a few magazines (available in print and on the web) that will enable inspectors and plan reviewers to keep abreast of the changing technologies.

IAEI News.	www.iaei.org/magazine
Solar Pro.	www.solarprofessional.com
Home Power.	www.homepower.com
Solar Today.	www.ases.org

02

PV Fundamentals and Calculations

To understand how and why the *Code* is different for PV systems compared with other electrical systems, it is necessary to first delve into the mysteries of what they are and how they operate. Inspectors need to be familiar with the unique characteristics of PV modules because the ampacity calculations and overcurrent device ratings don't follow the usual rules of thumb used in the electrical trades. Additionally, some PV installers may not be using the proper calculations either.

Current Sources

Photovoltaic modules are current sources of energy, as opposed to voltage sources such as 120-volt ac outlets in homes or offices or 12-volt dc batteries in cars. The output of a PV module, in terms of voltage and current, depends on the load placed on that module. The output is dc, and the load is typically supplied by a utility-interactive inverter, which changes the dc energy into ac energy.

Standard Test Conditions

Photovoltaic modules are rated for voltage and current output when exposed to a set of standard test conditions. The standard solar intensity (called *irradiance*) is set at 1000 watts per square meter (W/m^2). This is an international constant and is near the average value of irradiance at sea level on the surface of the earth. Modules are also rated at a standard module/cell temperature of 25 degrees Celsius (C) [77 degrees Fahrenheit (F)]. These two values of irradiance and temperature are referred to as standard test conditions (STCs). When the module is exposed to these standard test conditions (and connected to the correct load), the module will produce the rated power at a maximum power point voltage (V_{mpp}) and with a maximum power point current (I_{mpp}).

During evaluation of the safety of the module [performed by a nationally recognized testing laboratory (NRTL), leading to the module being listed as required by *NEC* 690.4(D)], the ratings of the module are also verified to be within some percentage of the label values. The tolerance on the label values is usually 10 percent but may be as low as 3 percent.

A PV module can be short-circuited indefinitely without damage. And, as will be shown in later chapters, the wiring, the switchgear and the overcurrent protection are designed in a way that will allow entire PV arrays to be short-circuited without damage.

Ambient Temperature

Although Section 310.15(B)(3)(c) has an Informational Note on a source for ambient temperatures, many people will just use the local weather station data.

> "Informational Note: One source for the ambient temperatures in various locations is the ASHRAE *Handbook—Fundamentals*."

The temperature adjustments for raceways in sunlight were substantial in the 2014 *NEC*, but the requirements were reduced in the 2017 *NEC*.

When addressing PV circuits on the roofs of buildings, consideration should be given that the local weather station may be at an airport outside the city and could be in an open area. The PV system may be installed in the city where the city heat bubble may increase ambient temperatures over the weather station's measurement. On a flat-roofed building with parapets, limited airflow may increase ambient temperatures on the roof even more. These increases are before the application of any solar heating on conduits and should be considered when considering the 50-year lifespan of a PV system in this environment.

What about the conductors and raceways behind those hot PV modules? If these conductors or raceways are within 3–4 inches of the backs of the PV modules, an ambient temperature of 75°C has been used in many parts of the country. If the conductors or raceways are spaced further away, then a 65°C ambient temperature may apply. Local situations may dictate different temperatures.

Current and Voltage Measurements — The I-V Curve

Measuring the module or array output under short circuit conditions will allow measurement of the short-circuit current (I_{SC}), which will be used in PV system sizing and in many *Code* calculations. A voltage measurement under short-circuit conditions will yield zero volts.

If a voltmeter is used to measure the voltage output of a PV module or array that is not connected to any load, the voltage obtained will be the open-circuit (no load) voltage (V_{oc}). A current measurement would be zero for this open-circuit condition.

If simultaneous voltage and current measurements are taken on a PV module or a PV array and these measurements are plotted for various loads, a graph that shows the electrical characteristics of a PV module could be shown. The graph would have current (I) on the vertical axis and voltage (V) on the horizontal axis. This graph or plot is shown in Figure 2.1 and is called an I-V curve. Similar curves appear in data sheets for PV modules.

The point to the right on the horizontal axis is the open-circuit voltage (V_{OC}) and the current at this point is zero. On the vertical current axis, the curve intersects the axis at the short-circuit current (I_{SC}) where the voltage is zero.

Power Equals Voltage Multiplied by Current

Each point on the I-V curve represents a value of voltage and a value of current at a particular load. Multiplying the voltage (V) by the current (I) will calculate the power (P) produced by a module and delivered to the load.

$$P = V x I$$

From this relationship, it can be seen that the module delivers no power at either the open-circuit voltage point or at the short-circuit current point because one of the factors of power is zero at these points.

However, if other points are examined on the curve (producing power from different loads), it will be noted that the power is not zero at these points. If the power output curve is added to the I-V curve, the graph shown in Figure 2.2 is obtained, which includes the I-V curve in blue and the power curve in red. The horizontal axis for the combined graph is still volts, but the vertical axis (on the right) for the power

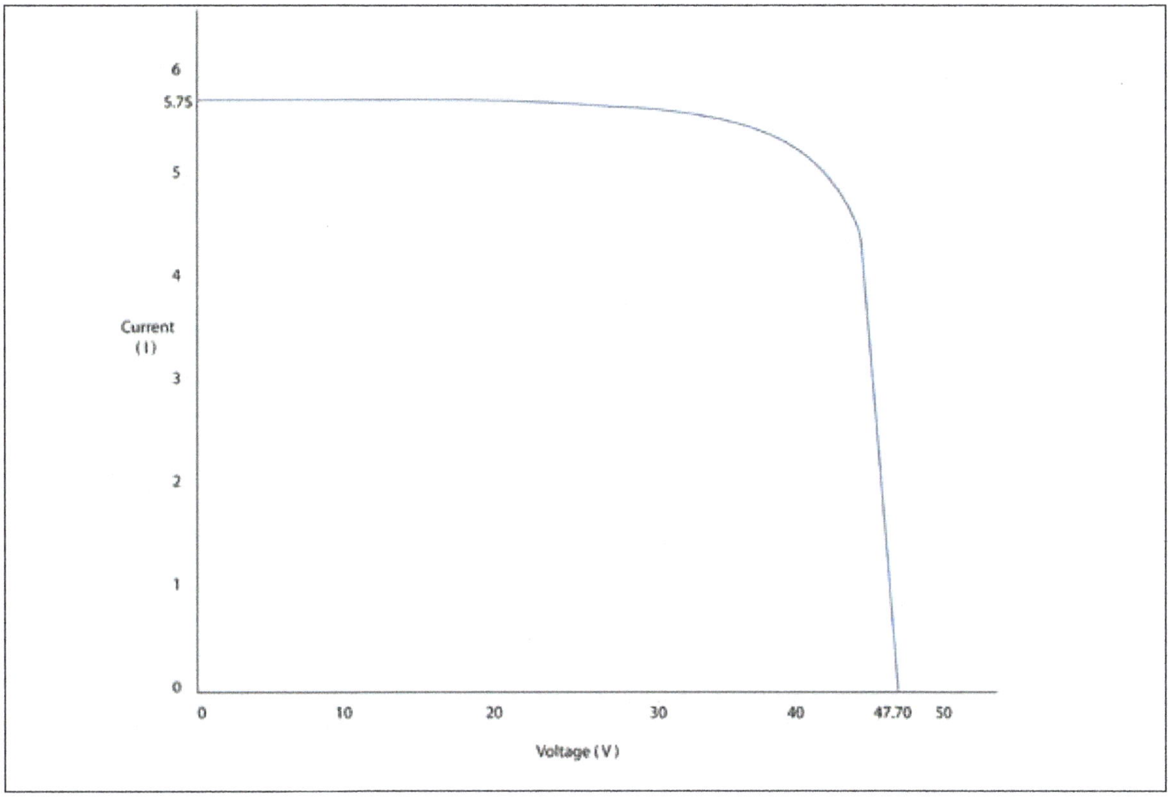

Figure 2.1 • I-V curve for a PV module.

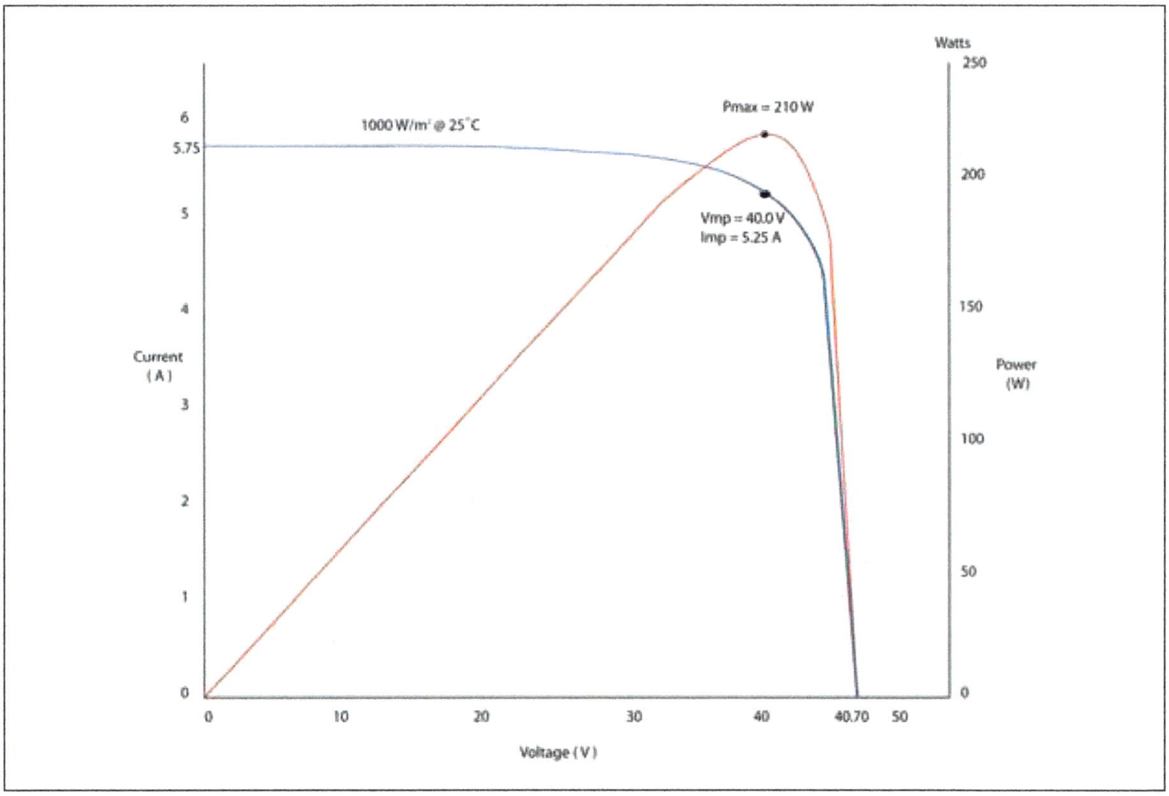

Figure 2.2 • I-V and power curve.

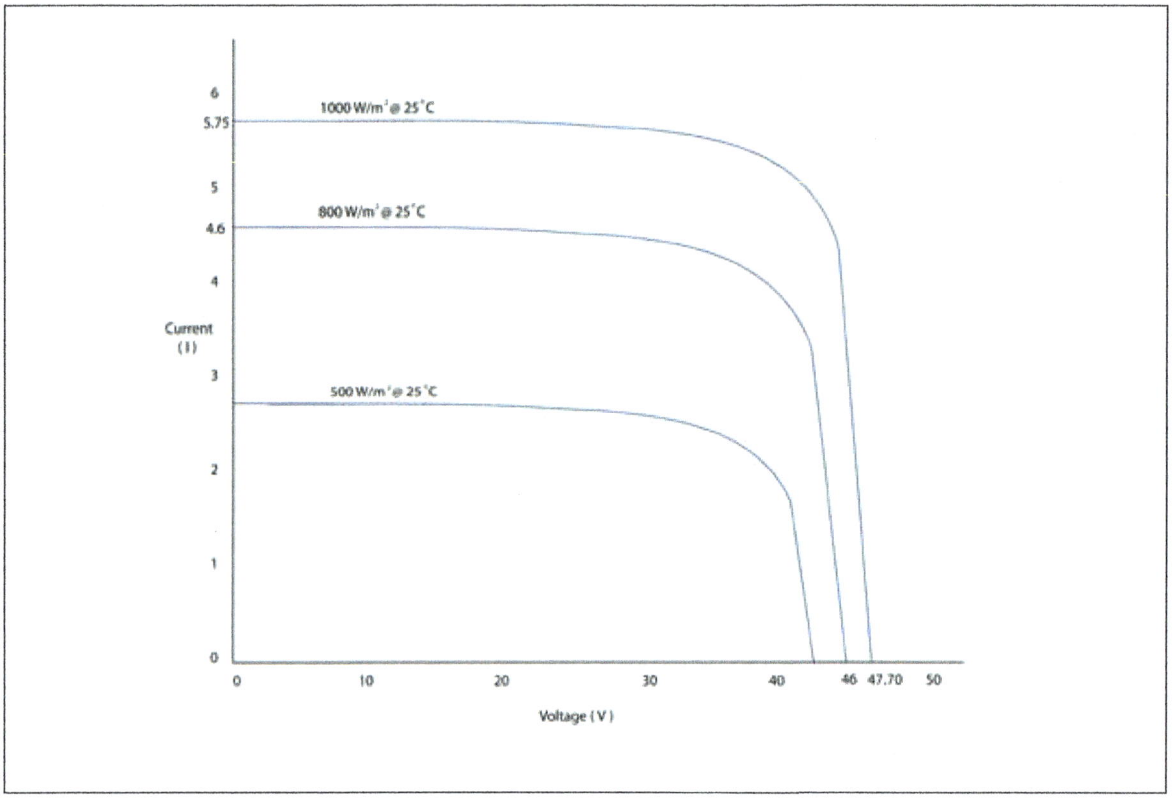

Figure 2.3 • I-V curve variations due to varying sunlight.

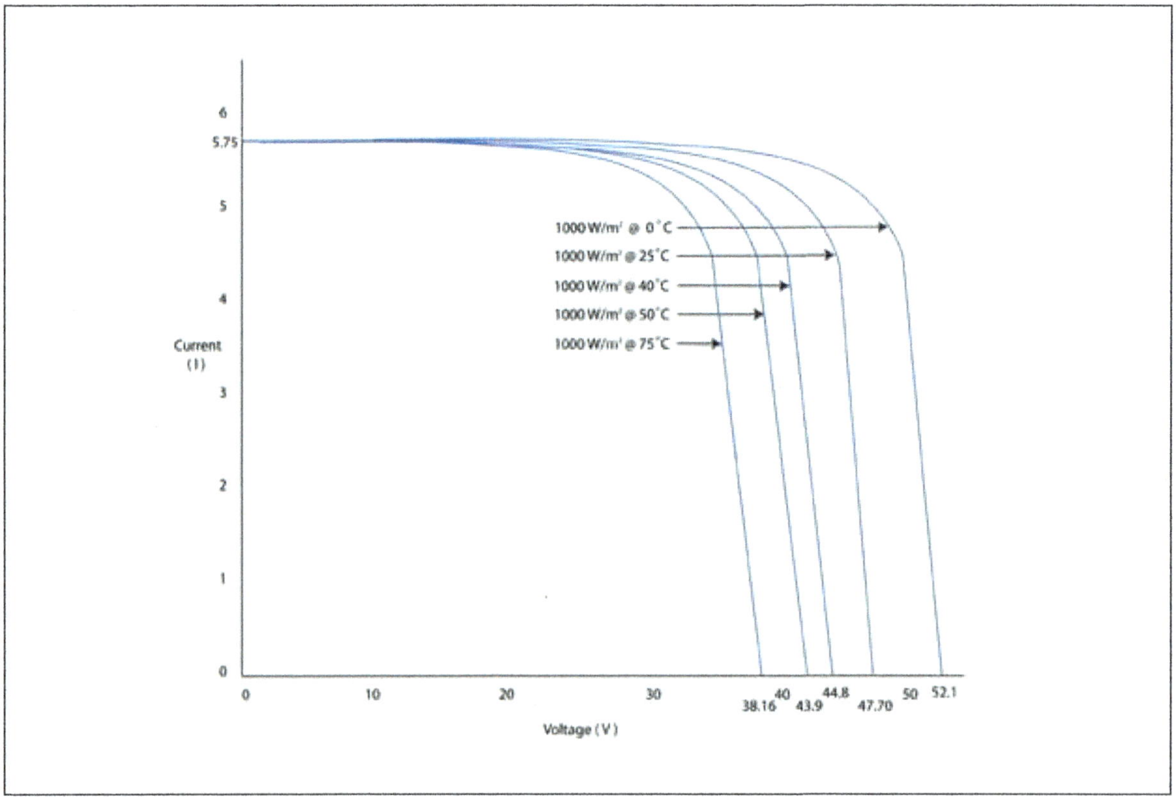

Figure 2.4 • I-V curve variations due to varying temperature.

curve is now marked watts. The curve for power shows that it reaches a peak for some load between I_{SC} and V_{OC} and this point is called the maximum power point (P_{mpp}). Associated with the power at the maximum power point are a maximum power point voltage (V_{mpp}) and a maximum power point current (I_{mpp}).

It should be noted that the output voltage of a PV module is not constant and varies with the load. This output is changed by several external environmental conditions in addition to the connected load.

Sunlight Produces Current

The current output of a PV module is directly proportional to the intensity of the sunlight falling on it. The rated currents (both I_{SC} and I_{mpp}) are output at the standard test-condition irradiance of 1000 W/m^2. However, PV modules are exposed to irradiance values from 0 (at night) to 1500 W/m^2 (if cloud, water, snow, or sand enhanced) and the current follows changes in the intensity of the sunlight. A 10% reduction in the irradiance value will result in a 10% reduction in I_{SC} and I_{mpp}. However, the open circuit voltage (V_{OC}) is relatively unchanged with variations in irradiance. Figure 2.3 shows the I-V curves for a PV module as the sunlight intensity varies from 1000 W/m^2 down to 500 W/m^2. As can be seen, the I_{SC} changes in direct proportion to changes in irradiance but V_{OC} and V_{mpp} do not vary nearly as much.

This is a significant fact. The voltage on a PV module or PV array will generally be present at very low levels of light, such as at dawn or dusk. Photovoltaic arrays can have hundreds of volts on the wiring at dawn and dusk even before the sun directly illuminates the front of the modules. Hazardous voltages on exposed terminals in dc combiners, disconnects, and input terminals to inverters will be present.

A second variation in module output and the I-V curve is caused by temperature. Module current is relatively insensitive to temperature, but both the V_{OC} and V_{mpp} voltages will be affected. In crystalline PV modules, V_{OC} varies inversely with temperature at approximately 0.5% per degree Celsius. Peak-power voltage (V_{mpp}) varies inversely at approximately 0.4% per degree Celsius. Figure 2.4 shows this relationship; as temperature goes down, voltage increases.

Turn Off the Light

Photovoltaic modules in utility-interactive systems are connected in series and the open-circuit voltage may approach 600 volts (dwelling), 1000 volts (commercial), and 1500 volts (utility-scale) in cold weather. The only way to effectively turn off all electricity from a PV module or a PV array is to cover it with an opaque material. Working at night on array wiring is an option, but worker safety would be a concern. Lightning in the distant sky has been known to illuminate arrays enough to produce electric shocks. High-intensity lighting used by fire services may also create unsafe voltages in PV arrays.

Determining the PV Size and Output to Match the Inverter

In the PV design process, the array output must be matched to the utility-interactive inverter input. As noted previously, the module voltage and current outputs are not constant. The system designer must first ensure that, in cold weather conditions, the array output voltage will not exceed the voltage rating of the inverter, the conductor insulation, or other connected equipment. This limit will frequently be as high as 600–1500 volts. In a similar manner, all conductors, overcurrent devices, and switchgear must be able to handle the array output current under worst-case conditions of high sunlight intensity. Exceeding the voltage rating on inverters, conductors, switchgear, or other equipment would be a *Code* violation and has been known to damage such equipment. [See *NEC* 110.3(B).]

Early pioneers in PV, including PV module manufacturers, NASA's Jet Propulsion Laboratories (space programs), Underwriters Laboratories, and the National Fire Protection Association, realized that because of these variations in the output of PV modules, special consideration would need to be made to handle these systems in the *National Electrical Code (NEC).*

UL Standard 1703, *Standard for Flat-Plate Photovoltaic Modules and Panels,* was written to establish safety requirements (mechanical and electrical) that PV modules would be required to meet. Until around 2012, requirements in module instruction manuals were written that modified the values of the standard test condition parameters printed on the back of modules.

These modifications dealt with the fact that sunlight intensity in many parts of the country could exceed the standard test condition of 1000 W/m^2 for three hours of more on numerous days throughout the year. Yes, PV modules can deliver more than rated current and rated voltage as environmental conditions vary.

Three hours per day represent continuous duty in the *Code.* It was decided in the early days of PV systems that *Code* calculations should be based on worst possible outputs and that those outputs would be considered to be continuous 24 hours per day, 7 days per week, 52 weeks per year and not vary on a daily cycle with the sun. Therefore, all voltages and currents used in calculations for PV systems in the *Code* are adjusted from standard test condition measurements in a way that ensures electrical systems meet *Code* requirements for safety and that all equipment will be operated within limits established by both the *Code* and UL standards.

Adjustments—Short-Circuit Current

The first adjustment is made to the module short-circuit current. Until May 2012, UL Standard 1703 required that module manufacturers have the following statement (or its equivalent) in instruction manuals that are shipped with every PV module.

"Multiply the short-circuit current (I_{SC}) marked on the back of the module by 125% before applying any requirements established by the *NEC.*"

This 125% factor is applied to the short-circuit current to ensure that the current used in any *Code* ampacity or overcurrent device rating calculations will never be exceeded by the actual module output under high-irradiance conditions. A 125% factor on I_{SC} would be equivalent to exposing the module to 1250 W/m^2 of irradiance, and that value of solar intensity has never been measured on a continuous basis.

Transient values of sunlight may reach 1500 W/m^2 due to cloud reflectivity enhancements and reflections off sand, water, and snow. However, these conditions are short-term because the clouds move, as does the sun (through earth rotation). These are not stable situations and last for only a few minutes.

Instructions for most flat-plate PV modules prohibit the installation of any permanent light-reflecting or light-concentrating surface near the modules.

Adjustments — Open Circuit Voltage

As noted previously, open-circuit voltage (V_{OC}) varies inversely with temperature. Instruction manuals for PV modules have the following statement or its equivalent:

"Multiply the open-circuit voltage (V_{OC}) marked on the back of the module by 125% before applying any requirements established by the *NEC.*"

For crystalline silicon, a factor of 125% on V_{OC}

represents the open-circuit voltage of a module at -40°C (-40°F). Temperatures lower than -40° are found in some parts of the country and further adjustments to V_{OC} may have to be made where these temperatures are expected.

NEC Leads UL Standard 1703

In 1996, during preparations for the *1999 NEC*, the PV industry, UL, and the National Fire Protection Association reached a joint agreement to place all module output correction factors in the *Code*. Those proposals were agreed to by all, and the *1999 NEC* was modified accordingly.

In Section 690.8, a requirement was established to multiply the module short-circuit current by 125%. This *duplicated* the module manual instruction to adjust the I_{sc} to account for normal and expected irradiance values above 1000 W/m^2.

Because many areas of the country never have temperatures as low as -40°, a temperature-dependent table was added to Section 690.7 showing multiplication factors for V_{oc} that vary from 1.00 at an expected low temperature of 25°C (77°F) to 1.25 at -40°. Again, instructions in the module manual duplicated these *Code* requirements.

Because newer module technologies have different temperature coefficients than crystalline silicon, the *2017 NEC* permits module manufacturers' temperature coefficients be used where available rather than Table 690.7.

UL modified Standard 1703 in 2012 to remove the 125% multipliers on V_{OC} and I_{SC} from PV module instruction manuals. This solved a significant problem for inspectors and installers who were required to follow *NEC* section 110.3(B), which states that all equipment be installed in accordance with included instructions. If followed, this section of the *Code* (and 690.7 and 690.8) create a double calculation with far too conservative results. Module instruction manuals have been modified to remove these two requirements and only the *Code* requirements will remain for PV modules being installed where the *NEC* is in force. Advocacy from electrical inspectors throughout the country inspired these changes.

The 125% Continuous Loads Factor

The requirement to use 125% of the continuous load (or 125% of the maximum current in PV circuits) has been with us for many years. It was reportedly added to the *Code* to address nuisance operation of overcurrent devices in enclosures when conductors were operated at 100% ampacity continuously. The temperature rise associated with I^2R losses in the conductors, and in the thermal overcurrent devices themselves, was causing some of these devices to overheat and nuisance trip or blow. Using the 125% factor results in a conductor that is not loaded continuously at more than 80% of rating. Note that 1.25 and 0.8 are mathematical reciprocals of each other (1/1.25 = 0.8 and 1/0.8 = 1.25), a fact that will be used below.

A Second 125%

Throughout the *NEC*, feeders and branch circuits (as well as overcurrent devices) are rated to operate on a continuous basis at no more than 80% of rating. Section 215.2(A)(1) has this requirement, which is also found in other sections of the *Code*:

> "**(1) General.** Feeder conductors shall have an ampacity not less than required to supply the load as calculated in Parts III, IV, and V of Article 220. Conductors shall be sized to carry not less than the larger of 215.2(A)(1) (a) or (b).
>
> (a) Where a feeder supplies continuous loads or any combination of continuous and non-continuous loads, the minimum feeder con-

ductor size shall have an allowable ampacity not less than the noncontinuous load plus 125 percent of the continuous load.

(b) The minimum feeder conductor size shall have an allowable ampacity not less than the maximum load to be served after the application of any adjustment or correction factors."

The intent of this code requirement, other than the obvious rating factors, is that the 125-percent factor (a) is not to be applied at the same time the "conditions of use" factors—such as temperature and conduit fill corrections on ampacity—in (b) are applied. From a math point of view 1.25 (125%) and 0.8 are reciprocals, so they can be used two ways. The conductor size can be calculated by taking 1.25 times the continuous load (*maximum current* for a PV or other generator) or by taking 0.8 times the conductor ampacity (before adjustments for conditions of use) to find the maximum continuous currents.

This same 125-percent factor (or the conditions-of-use factors if they result in requiring a larger conductor) is applied to PV source and output circuits, so the dc PV source circuit or dc output short-circuit current might be multiplied by 1.25 *twice* before selecting a conductor or overcurrent device. A factor of 1.56 (1.25 x 1.25 = 1.56) is commonly used to determine ampacity, but this is a shortcut and the slightly more complex actual calculations will be addressed in below.

The V_{OC} should be calculated following the requirements of 690.7. The factors will be less than 125 percent unless very low temperatures (-40°) are expected at the installation location. Table 690.7(A) may be used, but it will yield more conservative values for the cold weather maximum system voltage, which, in many cases, may not result in the most cost-effective system design.

Energized and Safe for Decades

It has been shown how module output varies with environmental conditions. The *Code*-required corrections to the rated output are used to ensure that system electrical equipment never has to handle more current or voltage than it was designed for. The rated output has been adjusted for worst-case conditions under the assumption that these worst-case conditions exist continually, even when it is known that output goes to zero every night. Photovoltaic modules will produce dangerous amounts of voltage and current for 40 years or more. These adjustments, and the other *Code* requirements, represent minimums to ensure that this sunlight-generated electricity is safely contained for that period of time or longer.

PV Math—Some Sample Calculations

Looking at the PV array in a PV system, many installers and inspectors are confused by new system voltage calculations that may be required by the *Code* specific to PV systems. *Code* Informational Notes also address voltage drop that may be applied to the dc wiring from the array to the inverter. This section will cover both of those subjects.

PV Math—Module Open-Circuit Voltage

A PV module, or a string of series-connected modules, has a rated open-circuit voltage that is measured (and labeled on the module) at an irradiance of 1000 W/m^2 and a cell temperature of 25°C (77°F). This voltage increases from the rated voltage as the temperature drops below 25°C. It is necessary to calculate this voltage at the expected lowest temperature at the installation location to ensure that it is less than the maximum input voltage of the inverter *and* less than the voltage rating of any connected conductors, switchgear, and overcurrent devices (usually 600 volts). Because parallel connections of strings do not affect open-circuit voltage, the number of strings connected in parallel is not involved with this calculation.

Where module temperature coefficients are available, Section 690.7 of the *NEC* allows the open-circuit voltage of a PV array to be determined at the lowest expected temperature at the installation location. Alternatively, Table 690.7(A) can be used to determine a multiplier that was applied to either the module- or string- (a series connection of PV modules) rated V_{OC}.

The rated V_{OC} is measured at 25°C (77°F) and is printed on the back of the module and in the module's technical literature. To use Table 690.7(A), determine the lowest expected temperature, look up the factor from the Table for that temperature (which ranges between 1.02 at 24°C to 1.25 at -40°C), and multiply the factor by the rated V_{OC}.

For example, a module has a V_{OC} of 35 volts (at STC) and is going to be installed where the temperature dips to -17°C. The factor from Table 690.7(A) is 1.18 and the cold temperature V_{OC} for this module is 35 x 1.18 = 41.3 volts.

If 12 modules were going to be connected in series, the string V_{OC} in cold weather would be 12 x 41.3 = 495.6 volts.

The string voltage at STC-rated conditions could also be calculated first and then apply the temperature factor applied. In this case, the 12 modules in series would have a string open-circuit voltage of 12 x 35 = 420 volts at 25°C. Then the 1.18 factor is applied to get 1.18 x 420 = 495.6 volts; the same answer as before.

While Table 690.7(A) is still valid and was refined with 5°C increments in the 2011 *NEC*, new modules may have different technologies than the silicon module technology used to develop the table. The use of the module temperature coefficients will provide a more accurate calculation of the cold weather maximum system voltage. To add to the confusion, PV module manufacturers present these temperature coefficients in two different ways.

Percentage Coefficients

One way of presenting these data is to specify them as a percentage change, and they are expressed as a percentage change in V_{OC} for a *change* in temperature measured in degrees Celsius. Note that the temperature used is a *change* in temperature from the rated 25°C.

For example: The V_{oc} temperature coefficient is given as -0.36% per deg Celsius, or -0.36% / °C.

The module has a V_{OC} of 45 volts at 25°C (77°F) and is going to be installed where the expected lowest temperature is -10°C (14°F). Because the temperature coefficient is given in degrees Celsius, all numbers must be in degrees Celsius. The *change* in temperature is from 25°C to -10°C. This represents a change in temperature of 35 degrees. The minus sign in the coefficient can be ignored if we remember that the voltage *increases* as the temperature goes *down* and vice versa. Of course, if you are an engineer or a mathematician, feel free to use the minus sign in an algebraic equation.

Applying the coefficient shows that the percentage change in V_{OC} resulting from this temperature change is 0.36% / °C x 35°C = 12.6%.

Note that where division and multiplication are involved in a calculation, they are performed from left to right. Where additions and subtractions are combined with multiplications and divisions, the additions and subtractions are performed before any multiplications. Parentheses may be added to clarify the order of the operations, and calculations inside the parentheses should be performed first.

This percentage change can now be applied to the rated V_{OC} of 45 volts. And, at -10°C, the V_{OC} will be 1.126 x 45 = 50.67 V.

For the mathematically oriented person, the equation looks like this:

V_{OC} [cold] = V_{OC} [STC] x (1 + ((-0.36)/100) x (-10-25))) = 45 x (1 + .0036 x 35) = 45 x 1.126 = 50.67

Eleven of these modules could be connected in series and the cold-weather voltage would be 11 x 50.67 = 557.37 V, and that voltage is less than a 600-volt equipment limitation.

Millivolt Coefficients

Other PV module manufacturers express the V_{oc} temperature coefficient as a millivolt coefficient. A millivolt is one one-thousandth of a volt, or 0.001 V.

A typical module with an open-circuit voltage (at 25°C) of 65 volts might have a temperature coefficient expressed as:

-240 mV per degree Celsius, or -240 mV/°C

If it is installed where the expected low temperature is -30°C (-22°F), then there is a 55°C degree change in temperature from 25°C to -30°C. Again, the calculations must be accomplished in degrees Celsius, ecasue that is the way the coefficient is presented.

Millivolts are converted to volts by dividing the millivolt number by 1000.

240 mV / 1000 mV/V = 0.24 volts

The module V_{OC} will increase 0.24 V/°C x 55°C = 13.2 volts as the temperature changes from 25°C to -30°C.

The module V_{OC} will increase from 65 volts at 25°C to 65 + 13.2 = 78.2 volts at the -30°C temperature.

In this PV system, the inverter maximum input voltage was listed as 550 volts. How many modules could be connected in series and not exceed this voltage? The maximum inverter voltage of 550 volts is divided by the cold-weather open-circuit voltage for the module of 78.2 volts.

550 / 78.2 = 7.03 modules and the correct answer would be seven modules.

7 x 78.2 V = 547.4 V

Eight modules could not be used because the open-circuit, cold-weather voltage would exceed 550 volts.

8 x 78.2 V = 625.6 V

Expected Lowest Temperature?

What is the source of the expected lowest temperature? Normally, this temperature occurs in the very early morning hours just before sunrise on cold winter mornings. The PV modules are, in many cases, a few degrees colder than the air temperature due to night-sky radiation effects. The illumination at dawn and dusk are sufficient to produce high V_{OC}, even when the sun is not shining directly on the PV array and has not produced solar heating of the modules. And in many locations, cold temperatures are coupled with high winds and the winds can remove solar heating from the module, even in bright sun.

A conservative approach would be to get weather data showing the record-low temperature of the area and use this as the expected low temperature. Other data show more moderate low temperatures associated with the data used to size heating systems. See the Informational Note on 690.7(A).

The National Renewable Energy Laboratory maintains data that shows the record lows for many locations in the United States (http://rredc.nrel.gov/solar/old_data/nsrdb/1961-1990/redbook/mon2/state.html).

Local airports and weather stations may have historical data on low temperatures.

Also, weather.com has data on file accessble by ZIP codes (www.weather.com/weather/climatology/monthly/zip code).

Photo 2.1 • DC-to-dc converter. Solar Edge Power Optimizer. *Courtesy Solar Edge.*

PV Math—Module Short-Circuit Current

In most silicon PV modules, the module short-circuit current does increase very slightly as temperature increases, but the increase is so small as to be negligible at normal module operating temperatures. It is normally ignored.

Informational Notes—Voltage Drop

In common, utility-interactive PV systems, PV arrays may operate from 50–60 volts up to near 600 volts, depending on the system design. With nominal, peak-power, and open-circuit voltages to deal with, installers and inspectors are sometimes in a quandary as to how to calculate voltage drops from PV arrays to the inverters.

A utility-interactive inverter will normally operate in a manner that keeps the array voltage near the peak-power voltage (also called the maximum power point). While this voltage can vary with temperature—and temperatures vary considerably—using the rated maximum power point voltage and maximum power point current of the modules results in the easiest method of calculating voltage drops.

A typical PV array may have a single string of ten modules in series connected to the inverter 200 feet away with 10 AWG USE-2/RHW-2 conductors. The maximum power point ($_{mpp}$)numbers for the module are:

$$V_{mpp} = 45V$$
$$I_{mpp} = 5.5 \text{ amps}$$

For a single string of 10 modules, the string maximum power point numbers are:

$V_{mpp} = 450V$
$I_{mpp} = 5.5$ amps

Table 8 in Chapter 9 of the *NEC* gives conductor resistance per 1000 feet at 75°C.

For an uncoated, stranded 10 AWG conductor, the resistance is 1.24 ohms per 1000 feet.

The total conductor length (both ways) must be used in the calculation and this is 400 feet.

The resistance for 400 feet of a 10 AWG conductor is 400/1000 x 1.24 = 0.496 ohms.

The current at the maximum power point is 5.5 amps. Voltage drop is found by multiplying this current by the conductor resistance:

5.5 x 0.496 = 2.728 volts.

Expressed as a percentage, 2.278/450 x 100 = 0.606% or about 0.6%, and that is much less than the Informational Note recommendation of 3% for most circuits. Of course, the losses in the PV dc disconnect were not counted, but they are typically less than 1% on these circuits.

Voltage Drop — Another Perspective

When dealing with the dc input circuits of a utility-interactive inverter, the inverter operates the array at the maximum power point with a maximum power voltage (V_{mpp}) and a maximum power current (I_{mpp}). Because these parameters are affected by irradiance and temperature, it is difficult to determine what voltage and current should be used in the voltage drop equation.

The procedure above uses module specifications at standard test conditions that are based on the rating conditions of an irradiance of 1000W/m² and a cell temperature of 25°C (which is not a very realistic temperature under actual operating conditions). One conservative approach is to use the module manufacturer's specified normal cell operating temperature from the module data when it is available. Normal cell operating temperatures are typically in the range of 42°C to 52°C and is a measured temperature of the module when the irradiance is 800 W/m², ambient temperature is 20°C and the wind is blowing across the module at 1m/sec. This temperature is used to adjust the V_{mpp} and I_{mpp} numbers presented at standard test conditions to values associated with the higher normal cell operating temperature. The maximum power voltage will be reduced and the maximum power current will be increased very little if at all. Voltage drop will be reduced.

Module-Level Power Electronics—When a PV Module Does Not Act Like a PV Module

Module-level power electronics (MLPEs) are found within PV module junction boxes and in dc-to-dc converters connected to the module output leads at the module (photo 2.1). Module-level power electronics decouple the standard test condition ratings of the PV module and the module output characteristics from the conductors leading to the rest of the string of modules and the PV array and the inverter. The calculations shown above dealing with the PV module characteristics (primarily voltage, current, and power), are no longer applicable to the output of a module with module-level power electronics.

Module-level power electronics perform several functions, which vary from product to product. Some will perform PV rapid shutdown system actions (690.12) using a remote signal. Others will peak-power track the individual module and provide an output that gets as much power from individual modules as possible given the environmental parameters. Still, other MPLEs will attempt to maximize the output of the module while also maximizing the output of the string containing that module. In nearly all cases, the specifications relating to the output parameters of the MLPE will not be comparable to the output of the basic PV module. In nearly all cases, the manufacturer's instructions provided with the listed or certified module-level power electronics device must be followed (110.3 B). There is no possible way that the *NEC* will be able to address these numerous varying operating parameters and installation requirements for the numerous systems in a general manner.

For example, when a certain dc-to-dc converter is connected to an appropriately sized module, the maximum voltage output of this converter is rated at 60 volts. Moreover, on the surface, it would appear that (at most) ten of these devices could be connected in series to a utility-interactive inverter with a dc input rated at a maximum of 600 volts. However, the manufacturer's literature suggests that, optionally, 15 of these devices could be connected to that 600-volt inverter—because the inverter from that manufacturer communicates with each of the dc-to-dc converters and holds its output under actual operating conditions to a total voltage per string of less than 600 volts.

While this particular dc-to-dc converter is connected to a PV module that has the normal PV module ratings on the label, they are to be essentially ignored and the ratings that apply to the dc-to-dc converter are to be used in determining whether it and the PV module have been installed correctly. Some dc-to-dc converters have a very low output (one or two volts) when they are disconnected from the inverter or from communications with the inverter. Of course, when the MLPE is fully contained within the module junction box, the certified or listed PV module with its internal electronics will have a label showing the output characteristics of the combination. The output characteristics of the module-level power electronics may, in many cases, no longer follow the classic temperature relationships of the output characteristics of a normal dc PV module.

Conductor Sizing and Overcurrent Device Ratings

Inspectors or plan reviewers should be very familiar with the methods of calculating currents, conductor sizes, and overcurrent device protection required in PV systems because, in many cases, PV installers will not have performed these calculation in a manner that follows *NEC* requirements.

Historically, most residential and light commercial electrical wiring and inspections of these systems have involved indoor wiring at room temperatures [30°C (86°F) or less]. The ampacity tables in *NEC* Section 310.15 and Table 310.15(B)(16) were developed with those conditions in mind. The commonly used molded-case circuit breaker is rated for use with conductors with 75°C insulation and they have a rated maximum operating temperature of 40°C.

With these conditions and equipment characteristics in mind, some electricians have generally used the 75°C insulated conductor ampacity tables in Table 310.15(B)(16) and not bothered too much with temperature corrections [Table 310.15(B)(2)(a)] and terminal temperature limits [110.14(C)] because they were not necessary or were included in the tables being used.

However, dc PV conductors normally operate in an environment that is too hot for conductors with 75°C insulation. Conductors with 90°C insulation must be used, and appropriate temperature and conduit fill corrections must be applied along with verification that connected equipment terminal temperatures (60°C or 75°C) are not exceeded. To do otherwise and use the short-cuts of the old days will result in conductors that may be larger than *Code* requirements (resulting in unnecessary costs) or that may have inadequate ampacities under the extreme conditions of use.

Throughout the *Code*, circuits are sized based on 125% of the continuous load plus the non-continuous load. See 210.19(A)(1) and 215.2(A)(1). This requirement establishes a situation where conductors and overcurrent devices are not subjected to more than 80% of rating. (Note: 1/1.25 = 0.80.)

Electricians typically use the 125% factor and then also apply the conditions of use factors (temperature and conduit fill) sequentially. In recent editions, the *NEC* has clarified the requirements in 210.19(A)(1) and 215.2(A)(1), as well as in 690.8, so that both factors are *not* to be applied at the same time to the same circuit. See the 125 percent requirement below.

In the *Code*, we have at least two or three requirements that must be met in sizing conductors.

First is the definition of *ampacity* found in Article 100. Ampacity is "The maximum current, in amperes, that a conductor can carry continuously under the conditions of use without exceeding its temperature rating."

Next is the 125 percent requirement, or conditions of use factors. From 690.8(B), Conductor Ampacity: (In part):

> **(B) Conductor Ampacity.** PV system currents shall be considered to be continuous. Circuit conductors shall be sized to carry not less than the larger of 690.8(B)(1) or (B)(2)....
>
> **"(1) Before Application of Adjustment and Correction Factors.** One hundred twenty-five percent of the maximum currents calculated in 690.8(A) before the application of adjustment and correction factors.
>
> *Exception: Circuits containing an assembly, together with its overcurrent device(s), that is listed for continuous operation at 100 percent of its rating shall be permitted to be used at 100 percent of its rating.*
>
> **(2) After Application of Adjustment and Correction Factors.** The maximum currents calculated in 690.8(A) after the application of adjustment and correction factors."

Note for the Exception: In other sections of the *Code,* such as 215.2(A)(1)(a), this exception is expanded to indicate that with this type of overcurrent protection, the conductor ampacity can be rated at 100% of the continuous currents (or maximum currents in the case of PV circuits). The author is not aware of any dc overcurrent protective device in an assembly rated for 100% operation.

> *215.2(A)(1)(a) Exception No. 1: If the assembly, including the overcurrent devices protecting the feeder(s), is listed for operation at 100 percent of its rating, the allowable ampacity of the feeder conductors shall be permitted to be not less than the sum of the continuous load plus the noncontinuous load.*

Author's Note: This book will not address the 2017 *NEC* requirements related to "adjustable electronic overcurrent protective devices" found in 690.8(B), 690.8(B)(3), and 690.9(B)(3) because these devices and their requirements are not well defined in either the 2017 *NEC* or in the relevant UL Standards. At best, they are adjustable trip circuit breakers and are handled like other circuit breakers. Revisions, clarifications, or deletions to these requirements are expected in the 2020 *NEC.*

Terminal Temperature Limitations and Operating Temperature Limitations

Section 110.14(C) requires that the temperature of the conductor in actual operation not exceed the temperature rating of terminals on the connected equipment.

An added requirement for any listed equipment, such as overcurrent devices, is that they *not* be used in a manner that deviates from the listing or labeling on the product [110.3(B)]. Most PV combiners operating outdoors in the sunlight will have internal temperatures that exceed the 40°C rated operating temperatures of commonly used fuses and circuit breakers. Overcurrent devices listed for (and required in) PV applications will have a 50°C-rated operating temperature.

Calculating Conductor Sizes and Overcurrent Device Ratings

The following method for determining ampacity meets the three *Code* requirements above and finds the smallest conductor that can be used to meet these requirements. The method also determines the rating of the overcurrent device where required. It is consistent with the requirements found in Article 690 and elsewhere in the *Code.*

Step 1. Determine the maximum circuit current [690.8(A)]

Photovoltaic dc circuits and PV ac circuits are not "load" circuits; the *Code* uses the term *current* instead of *load.* For *Code* calculations, all dc and ac PV currents are considered continuous and are based on worst-case outputs or are based on safety factors applied to rated outputs [690.8(B)]. The term *maximum* is used in Article 690 instead of *continuous* current used elsewhere in the *Code.* Daily variations in these currents is ignored.

A. Photovoltaic dc circuits

In the dc PV source and dc PV output circuits, maximum currents are defined as 1.25 times the rated short-circuit current (I_{SC}) marked on the back of the module. If a module had an I_{SC} of 7.5 amps, the maximum current would be 7.5 x 1.25 = 9.375 amps [690.8(A)(1)].

If three strings of modules (module I_{SC} = 8.1 amps) were connected in parallel through a fused dc combiner, the PV output circuit of the combiner would have an I_{SC} of 3 x 8.1 = 24.3 amps [690.8(A)(2)] and the maximum current would be 24.3 x 1.25 = 30.375 amps [690.8(A)(2)].

An exception to these calculations is allowed for PV systems of 100 kW or greater, where a professional electrical engineer may make alternate maximum current calculations using an industry accepted method.

B. AC inverter output circuits

In the ac output circuits of a utility-interactive inverter (or in the ac output circuit of a stand-alone inverter), the continuous current is taken at the full rated output power of the inverter. It is *not* measured at the actual operating current of the inverter (which may be a small fraction of the rated current due to a small PV array connected to a large inverter). Usually the rated current is at the nominal output voltage (120, 208, 240, 277, or 480 volts). The rated output current is usually specified in the manual, but may be calculated by dividing the rated power by the nominal voltage. For stand-alone inverters, which can provide some degree of surge current, it is the rated power that can be delivered continuously for three hours or more [690.8(A)(3)].

In some cases, the inverter specifications will give a rated current that is higher than the rated power divided by the nominal voltage. In that situation, the higher current should be used. This higher current usually has been determined at a lower-than-normal line voltage.

For a utility-interactive inverter operating at a nominal voltage of 240 volts and a rated power of 2500 watts, the continuous current would be 2500 W/240 V = 10.4 A.

A stand-alone inverter with a model number of 3500XPLUS operates at 120 volts and can surge to 3500 watts for 60 minutes. However, it can only deliver 3000 watts continuously for three hours or more. The rated output current would be 3000 W / 120 V = 25 A

C. Stand-alone inverter battery currents

In either a stand-alone system or a battery-backed-up utility-interactive system, the currents between a battery and an inverter must be based on the rated output power of the inverter (continuous for three hours or more) at the lowest input battery voltage that can provide that output power [690.8(A)(4)].

This current is usually marked on the inverter or found in the specifications, and that number should be used where available.

When not available, the input current can be calculated by taking the rated output power, dividing it by the lowest battery voltage that can sustain that power, and also by dividing by the inverter dc-to-ac conversion efficiency at that battery voltage and power level. For example: A 4000-watt inverter can operate at that power with a 44-volt battery input voltage and has a dc-to-ac conversion efficiency (inverting mode) of 85 percent. The dc continuous current will be 4000 W / 44 V / 0.85 = 107 A

On single-phase stand-alone inverters, the dc input current is rarely smooth and will usually have 120 Hz ripple current (root mean square [RMS]) that is larger than the calculated continuous current. The inverter's technical specifications should list the greatest continuous current.

Step 2. Calculate a conductor size [690.8(B)].

The conductor selected for any circuit must meet the 125 percent requirement or the conditions of use requirement. The correctly sized conductor is the larger of A or B below.

***Example 1*:** Three conductors are in a conduit in a boiler room where the temperature is 40°C. The maximum current in all three conductors is 50 amps. A copper 90°C insulated conductor is specified.

A. 125 percent requirement

The conductor must have an ampacity of 125 percent of the maximum current established in Step 1.

In our example, 1.25 x 50 = 62.5 amps, and this would indicate a 6 AWG conductor.

B. Conditions-of-use requirement

The conductor, after corrections for conditions of use, must have an ampacity equal to or greater than the continuous current found in Step 1.

Temperature correction factor = 0.91
Conduit fill correction factor = 1.0

Required ampacity at 40°C is 50 / 0.91 / 1.0 = 54.9 amps, and this would require an 8 AWG conductor.

The 6 AWG conductor is the larger of the two and is required.

***Example 2*:** Now there are six conductors in the conduit and the temperature has increased to 50°C. The maximum current is still 50 amps.

Temperature correction factor = 0.82
Conduit fill factor = 0.8

Conditions-of-use requirement: 50 / 0.8 / 0.82 = 76.2 amps, and a 4 AWG conductor is needed.

125-percent rule: 1.25 x 50 = 62.5 amps calling for a 6 AWG conductor.

The 4 AWG conductor is the larger of the two and must be used.

Step 3. Terminal temperature limits[110.14(C)]

The terminal temperature limits marked on the equipment must be used. If no temperatures are marked, then a 60°C limit is used for circuits rated at 100 amps or less or conductors 14-1 AWG. For circuits rated greater than 100 amps and for conductors greater than 1 AWG, a 75°C terminal temperature limit will be used [110.14(C)].

The following method is a terminal temperature estimation method and is *not* an ampacity calculation method. It is used after the conductor size has been selected based on the ampacity calculation.

Take the conductor size in Step 3 above. Find the lowest terminal temperature limit for this conductor at any termination. Use that terminal temperature limit (either 60°C or 75°C) to enter the ampacity per Table 310.15(B)(16). For the conductor size selected, read out the current in the correct column, either the 60°C column or the 75°C column. There are no temperature adjustments or conduit fill adjustments in this estimation process.

The current from the table must be equal to or greater than 125 percent of the maximum current. If the conductor meets this requirement, then the terminal temperatures are going to be less than the 60°C or 75°C limit for that conductor and that maximum current. The 125-percent factor accounts for many items not calculated in this simplified temperature *estimation* process.

Example 3. Take the 6 AWG conductor and 50 amps of maximum current used in Example 1 above. This conductor is connected to a terminal with a 60°C marking.

From Table 310.15(B)(16), a 6 AWG conductor in the 60°C column can carry a current of 55 amps.

1.25 x 50 = 62.5 amps. This is larger than the 55 amps from the Table and this terminal will be heated above 60°C.

When we increase the conductor size to 4 AWG, the table gives us 70 amps, which is greater than 62.5 amps and the terminal will stay below 60°C.

Example 4. Use the 4 AWG conductor selected in Example 2 connected to a terminal with a 75°C temperature limit. The maximum current is 50 amps.

1.25 x 50 = 62.5 A

A 4 AWG conductor in the 75°C column of

Photo 2.2 • Multiple small raceways routed across the building.

Table 310.15(B)(16) shows a current of 85 amps. Because this is greater than 62.5 amps, the conductor will operate cooler than the 75°C terminal temperature limit. No increase in conductor size is necessary.

An alternate, more conservative approach to terminal temperature limitations would be to use an ampacity table of 60°C or 75°C for the conductor size calculations. However, this will result in a conductor larger than necessary because the advantages of the 90°C conductor insulation are not considered.

Step 4. Calculate the rating of the overcurrent device, where required.

Because PV modules are current-limited, overcurrent devices are frequently not needed for one or two strings of PV modules connected in parallel [690.9(A) Exception]. In systems with three or more strings of modules connected in parallel, overcurrent devices are usually required.

A. Rating determined from maximum currents

The overcurrent device rating is determined by taking the maximum current for any of the circuits listed in Step 1 and increasing that maximum current by 125% (or by multiplying it by 1.25). Nonstandard values should be rounded up in most cases [690.9B)].

In a few rare cases, an overcurrent device *installed in an enclosure* or in an assembly may be tested, certified, and listed *as an assembly* for operation at 100% of rating. The conductor ampacity, before conditions of use are considered, can be 100% of the maximum current. The author knows of no dc overcurrent devices installed in an enclosure for PV systems that have such a rating.

B. Operating temperature affects overcurrent device rating

Standard overcurrent devices are listed for a maximum operating temperature of 40°C (104°F), and overcurrent devices listed for PV applications are rated for use at 50°C. Photovoltaic combiner boxes operating in outdoor environments may experience ambient temperatures as high as 50°C. Exposed to sunlight, internal temperatures may reach or exceed 55°C to 60°C. Anytime the operating temperature of the overcurrent device exceeds 50°C, it may be subject to nuisance trips at current values lower than its rating. In these situations, the manufacturer must be consulted to determine an appropriate derating. At high operating temperatures, an overcurrent device with a higher rating will activate at the desired current; however, using a manufacturer's certification of a rating at a higher-than-listed temperature may invalidate the listing on the device.

Step 5. Verify that the overcurrent device protects the conductor selected under the conditions of use

Where an overcurrent device is required, it must protect the conductor under conditions of use. Conductors may be protected using the round-up allowance found in 240.4(B).

Example 5. A circuit has a maximum current of 70 amps. After conditions of use (4 conductors in the conduit, 48°C) are applied, a 3 AWG, 90°C conductor is selected to meet all ampacity and 75°C terminal-temperature requirements.

The ampacity after conditions of use have been applied is 115 x 0.8 x 0.82 = 75.4 A.

The required minimum overcurrent device for this level of maximum current is 70 x 1.25 = 87.5 A.

A 90-amp overcurrent device would typically be used. It's been suggested to use an 80-amp overcurrent device, but that would result in running it at more than 80% of rating and, in dc PV circuits, could result in nuisance trips during short periods of cloud-enhanced irradiance.

However, the largest overcurrent device that could be used to protect the 3 AWG conductor

Photo 2.3 • Raceways spaced at varying distances from the roof.

with an ampacity of 75.4 amps is an 80-amp overcurrent device. A 90-amp overcurrent device is the smallest allowed in this circuit.

The conductor size would have to be increased to 2 AWG for full compliance with *NEC* requirements.

The ampacity of a 2 AWG, 90°C conductor under the conditions of use is 130 x 0.8 x 0.82 = 85.28 A. With the allowed roundup, the next standard value of fuse is 90 amps.

The 2 AWG conductor can be protected by the required 90-amp overcurrent device.

Raceways in Sunlight and Varying Conditions of Use

Note that 310.15(B)(3)(c) no longer requires all the temperature adders required by the 2014 *NEC*. Only conductors and raceways with bottoms less than 23 mm (7/8 in) from the roof are subject to a temperature adder of 33°C (60°F). Table 310.15(B)(3)(c) has been removed from the 2017 *NEC*.

Photovoltaic professionals may elect to continue to use Table 310.15(B)(3)(c) from the 2014 *NEC* (again, removed in the 2017 *NEC*) as a safety factor to increase the durability of the PV source and output circuits over the 50-plus-year life of the PV system. This is based on the fact that conduits in sunlight that are 12 inches above the roof, for example, have been observed to be too hot to handle and are certainly significantly hotter than the surrounding ambient air temperatures.

Suppose, for example, there is a large PV array being installed on a flat-roof commercial building. Each PV source circuit (aka string of modules) requires two conductors (positive and negative) plus an equipment grounding conductor that will originate at the end of each module string and be routed to a dc combiner

Table 310.15(B)(3)(a) Adjustment Factors for More Than Three Current-Carrying Conductors

Number of Conductors[1]	Percent of Values in Table 310.15(B)(16) through Table 310.15(B)(19) as Adjusted for Ambient Temperature if Necessary
4–6	80
7–9	70
10–20	50
21–30	45
31–40	40
41 and above	35

Table 2.1 • Table 310.15(B)(3)(a).

inside the building (out of the high-temperature, sunlit rooftop location). Each of these three-conductor circuits is installed in an EMT raceway and 12 such circuits are routed from the 12 module locations to a centrally located junction/pull box where all 12 circuits are routed into a single large EMT raceway with one equipment-grounding conductor (photo 2.2). That large raceway, now with 24 circuit conductors and one equipment-grounding conductor, is routed across the roof and then down into the building where it terminates in the dc combiner.

As the smaller and larger EMT raceways run across the roof, they are all subjected to a number of different environmental conditions. In some locations, they are in the shade and are subject to the ambient temperature. At other locations, they are in the sun and may be 1/4 inch from the roof, or 5 inches from the roof on standoffs, or even 24 inches above the roof as they cross over HVAC ductwork (photo 2.3). Each of these environmental conditions must be addressed and will be coupled with the conduit fill factors that vary with each of these circuits in their run from the modules on the roof to the dc combiner.

It is noted that XHHW-2 conductors are exempt from any temperature adders; this is a superior type of conductor for PV installations [310.15(B)(3)(c) Exception]. However, engineers may wonder why the XHHW-2 conductor has only a 90°C rating when it can be exposed to higher temperatures without damage.

Looking at the temperature adders in Table 310.15(B)(3)(c) (2014 *NEC*) and the conduit fill factors in Table 310.15(B)(3)(a) (2017 *NEC*) will allow us to make the calculations.

In examining these two Tables, it will be noted that the larger EMT raceway with an adjustment factor of 0.45 for the 24 circuit conductors in the shade is not as significantly affected as the smaller EMT conduit with no conduit fill adjustment, where the smaller conduit is only 1/4 inch from the roof and the adder is 33°C (2104 *NEC* and 2017 *NEC*).

The PV source circuit conductors are typically 10 AWG USE-2 or PV wire. At an ambient temperature of 51°C (estimated roof ambient temperature), the temperature correction factor is 0.76 from Table 310.15(B)(2)(a). Adding 33°C increases the temperature to 84°C and yields a correction factor of 0.29. This reduction yields a corrected ampacity for 0.29 x 40 = 11.6 amps. This reduction factor, due to the conduit being close to the roof, is more significant than the ampacity reduction due to conduit fill alone in the larger EMT, which yields .45 x 40 = 18 amps.

However, the large EMT may be in the sun and would be subject to both conduit fill and solar heating adjustments.

Looking at the larger EMT in the sun and 5 inches from the roof gives us a conduit fill correction of 0.45. And with a temperature increase from 51 + 17 = 68°C, yields an additional correction factor of 0.58. Multiplying the two factors together gives us 0.45 x 0.58 = 0.261; this is a more restrictive adjustment factor than the ½-in. conduit mounted against the roof and the ampacity is reduced (40 x .261 = 10.4 amps).

It would appear that while all different conditions of use should be examined, the physics of the situation indicates that attention should be focused on the larger raceways in the sun containing more conductors. Plan reviewers could easily do these calculations, but, unfortunately, this level of detail on conduit mounting is not frequently found in plans submitted with permit applications.

Plan reviewers, inspectors, and installers take note: In PV circuits, that electrician's old rule of thumb of using a 30-amp fuse or breaker on a 10 AWG conductor is frequently not valid. Also, the fusing in dc combiners should be carefully calculated along with the size of these source circuit conductors, as the PV module short-circuit current continues to increase as larger and larger PV modules are being built.

The 10-Percent, 10-Foot Rule

Section 310.15(A)(2) may give some relief for the high temperature conditions on the roof.

> "**Selection of Ampacity.** Where more than one ampacity applies for a given circuit length, the lowest value shall be used.
>
> **Exception:** Where different ampacities apply to portions of a circuit, the higher ampacity shall be permitted to be used if the total portion(s) of the circuit with the lower ampacity does not exceed the lesser 3.0 m (10 ft) or 10 percent of the total circuit."

On a roof with PV circuits, there are going to be several sections of the circuit that will have different calculated ampacities. If the worst-case (lowest-ampacity) section is less than 3.0 m (10 feet) in length, or less than 10 percent of the length of the adjacent section with a higher ampacity (whichever is less), then the higher ampacity may be used for both sections. The *Code* uses the term *circuit portion* and the example in the 2014 *NEC Handbook* uses a single hot section and then uses the entire circuit length to make the determination. In the PV environment, it appears reasonable to use just the length of the adjacent circuit with a higher ampacity, as this section will be the heat sink for the shorter conductor.

If the most restrictive circuit section can be eliminated as the determining ampacity for the whole circuit, then the next most restrictive section should be examined in the same manner. Eventually, considering all conditions of use, an ampacity of the circuit will be determined. At that point, other factors, such as terminal temperature limits, should be considered.

Sources of Fault Current

In most ac power circuits, the utility source of energy becomes the source of the overload or fault currents, and the current in a circuit usually flows from the utility source to the load. The requirements in Section 240.21 can be followed where circuits are protected from overcurrents where they receive their supply: the utility end of the circuit.

However, in dc PV circuits, the PV modules have limited current even when short-circuited under fault conditions. This limited current also applies to the strings of series-connected modules and, to some extent, the subarray PV output circuits where several strings of modules are connected in parallel. As we move from the modules, to the strings and to the output of parallel connected strings forming subarrays to the output of parallel connected subarrays, into the output of the entire array, the available currents that could contribute to faults and overloads increase. As these currents increase, the circuit conductors are increased in size appropriately and the ratings of overcurrent devices are also increased.

The source of potentially high overload currents and fault currents is not the PV module or the string of PV modules. It is the combined output of those strings in parallel and the combined output of subarrays connected in parallel that pose the source of overload and fault currents. Section 690.9(A) addresses this peculiarity of current-limited sources.

> "**Circuits and Equipment.** PV system dc circuit and inverter output conductors and equipment shall be protected against overcurrent. Overcurrent protective devices shall not be required for circuits with sufficient ampacity for the highest available current. Circuits connected to current limited supplies (e.g., PV modules, dc-to-dc converters, interactive inverter output circuits) and also connected to sources having higher current availability (e.g., parallel strings of modules, utility power) shall be protected at the higher current source connection."

With current-limited sources, such as PV modules and the ac output of utility interactive inverters, the location of the overcurrent device for the circuits is going to be located at the source that has the highest available fault current: the parallel-connected strings of modules in the dc circuits or the utility for the ac circuits.

03

PV Modules — Installation Considerations

Photovoltaic modules must be installed and connected in a manner that meets *Code* requirements and that will ensure a safe, durable system with maximum output for possibly 50 years or more.

The PV Array — Mechanical Considerations

The PV array consists of individual PV modules attached to a metal rack. That rack is usually attached to the structural members of the roof in a typical rooftop-mounted residential utility-interactive PV system. Although not an electrical-code issue, some attention must be given to the attachment of the PV array to the building structure. Various national, state, and local building codes address the attachment of structures to buildings—especially in earthquake and high-wind areas.

Most roofs in recent years have been built using span tables in the building codes or using trusses designed by professional engineers. Photovoltaic arrays may add up to 4 to 5 pounds per square foot of dead weight to the roof structural members, and that weight will be concentrated through the rack mounting feet. Also, because the PV arrays are mounted above the roof some distance (from zero to six inches or more), the roof may be subjected to both uplift and down-force wind loadings—again concentrated through the mounting feet of the rack. If the roof has several layers of old shingles under the array, the structural limit of the roof may be approached. Leaving up to two layers of old shingles in place is a common practice during re-roofing, so we can assume that the basic roofing structure has a safety factor allowing the extra load of old shingles *or* the PV array, but possibly not fpr several

Photo 3.1 • Array rack attachment point—used in dry climates. Without flashing, the roof may leak during heavy rain.

Photo 3.2 • Module with attached cables and connectors.

layers of shingles *and* a PV array.

Array racks must be attached to the structural elements of a roof (trusses or rafters). This will require penetrating the roofing surface material in a manner that is weatherproof for the life of the PV array or for the life of the roof—whichever is shorter (photo 3.1).

Stainless steel hardware is commonly used to connect modules to racks. Galvanized hardware is frequently used to bolt racks to the roofs. In both cases, corrosion resistance is a must in most climates.

The PV Array—Cable Types and Installation

Electrically, PV arrays consists of PV modules connected in series using permanently attached, exposed single-conductor cables with "finger safe" connectors (photo 3.2). The conductors are typically USE-2 or PV wire as allowed by *NEC* 690.31. Either conductor type is suitable for exposed dc PV module wiring. Photovoltaic wire is a "super" USE-2 that has a thicker jacket (the conduit fill tables cannot be used); passes a 720-hour accelerated UV test (and is marked *Sunlight Resistant*); and has the flame and smoke retardants of RHW-2. It can be used under and within the PV array for the module interconnections and in raceways in other locations. This cable is used on all modules because it is required on modules that can be installed both in the United States and in other countries.

Photo 3.3 • Protective barrier "guarding" PV conductors.

Although electrical connectors attached to the ends of module cables are "finger safe" when new, if they are opened under load, the dc arc may damage the insulation and the connectors may then pose a shock hazard. Therefore, requirements in 690.33 establish the need for locking connectors. A tool will be required to open these locking connectors. They will soon be appearing on most (if not all) PV modules, although they are only required when PV array wiring is operating above 30 volts and is readily accessible (690.33).

Another requirement that applies to readily accessible PV source and output circuit conductors operating at over 30 volts is found in 690.31(A). These conductors must be installed in type MC cable or guarded raceways. Unfortunately, most PV modules do not have junction

Photo 3.4 • Modules in landscape orientation.

Photo 3.5 • Modules in portrait orientation.

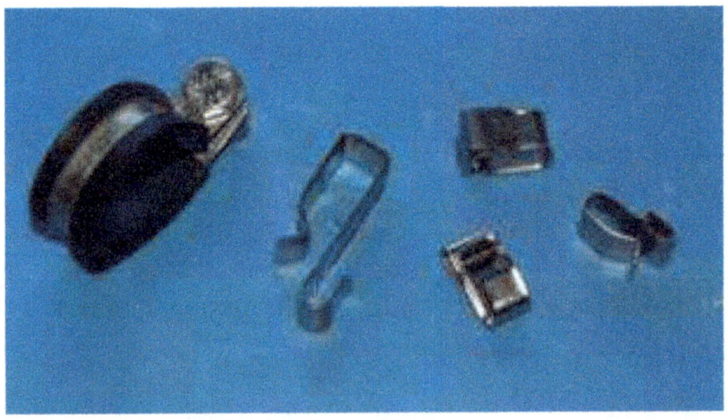

Photo 3.6 • Pipe clamps (EDPM and stainless steel) used to secure bundles of module conductors and special stainless steel clips to secure individual module conductors.

boxes with knockouts that would accept a raceway or MC cable. They come with permanently attached exposed, single-conductor cables and connectors with no provision for attaching a conduit or other raceway. Fortunately, most residential rooftop PV arrays are not readily accessible.

The solution, as noted by the word "guarded," is to make this wiring *not* readily accessible by placing a barrier behind the modules that prevents the wiring from being touched without removing the barrier. Fences with locked gates may not be a solution, because a basic maintenance requirement for a readily accessible ground-mounted PV array is keeping the grass mowed—a task usually done by people not qualified to be near PV or other electrical systems. Various barriers are on the market that either attach to the back of the PV module/rack assembly or to the rack to prevent ready access to exposed conductors (photo 3.3).

Conductor leads attached to modules are 40 inches long (or longer) to allow the series connection of modules when they are mounted in a landscape orientation (photo 3.4). When the modules are mounted in portrait orientation (photo 3.5), the excess lengths of conductors must be securely fastened against the module racks to resist abrasive damage due to wind, sleet, and ice. Many installers use plastic cable ties, but unless they are of very high quality, they may not last the required 40 years or more when exposed to the extremes of heat and ultraviolet radiation from sunlight. Some installers use a stainless-steel pipe clamp (loop strap) with an EDPM insert. There are numerous specially designed metal clips for this requirement (photo 3.6).

Cables and Connectors for PV Modules

Exposed cables and connectors used in PV source circuits are some of the most critical components of a PV system in terms of maintaining system durability and safety. These components are exposed to the extremes of weather for the life

Photo 3.7 • PV module connectors (normally not load-break rated).

of the system and may be required to be durable and safe longer than the system will produce power. Photovoltaic modules have a warranted lifespan for power output of 25 years, but they may produce dangerous amounts of voltage and current for as long as 40 or 50 years. In many cases, there will be a functioning (or more likely, nonfunctioning) inverter attached to a PV array output for that length of time. The quality of the components used, and the care taken in complying with the *Code*, determine the durability and ultimate safety of these cables and connectors.

Connectors for PV Modules and Related Equipment

Recognized Components

Photovoltaic modules come with connectors attached to the ends of the cables that have been permanently attached to the PV module (photo 3.7). There are multiple connector manufacturers supplying connectors to module manufacturers, and modules with a variety of connectors are being used. These connectors, as separate parts, are recognized components because they cannot meet the more stringent requirements of a fully listed connector, which would require that the connector to be capable of being opened under load, or opened at full current and full rated voltage, as well as meeting other requirements. An example of a listed, load-break rated connector assembly would be the ac plug used on appliances

Photo 3.8 • Microinverter with ac and dc connectors listed with the microinverter as isolating means. *Courtesy of Enphase Energy.*

and the 120-V ac receptacle in houses. Photovoltaic connectors, if they were to be listed, would have to be able to safely open a circuit at 600 volts or even 1000 volts dc at the rated current of the module, which could be (in some cases) as high as 10 amps or 12 amps. Interrupting direct currents is considerably more difficult than interrupting alternating currents and, at this time, none of these PV connectors are rated for load break operation at full voltage.

When these connectors are installed on a PV module or other PV component, the connectors are evaluated with that component and are certified as suitable for that application with that module or equipment. This certification still does not allow them to be operated under load, and nearly all of them will be marked "Not for Load Break Operation" or similar wording.

The exception to this recognized component rating relates to the connectors on some microinverters that have been listed with the inverter as a disconnecting means when used at the lower voltages and less strenuous operating conditions of these devices. These connectors meet the requirements of 690.15 as isolating devices (photo 3.8).

Compatibility Issues

Some connector manufacturers are advertising

that some of their connectors are interchangeable with connectors from other manufacturers in terms of an electrical and mechanical compatibility. On casual inspection by the AHJ and PV installer, this mechanical and electrical compatibility appears to be true. Unfortunately, Underwriters Laboratories and the members of the Standards Technical Panel for UL Standard 6703 for connectors have established that none of these mixed pairs of connectors have been evaluated for compatibility by an independent third-party Nationally Recognized Testing Laboratory like UL, CSA, TUV and ETL—the four OSHA-recognized NRTLs authorized to certify PV equipment. Each connector manufacturer modifies the materials and procedures used to manufacture their connectors in a proprietary manner. Even though there may be an electrical and mechanical compatibility at one point in time, there is no continued evaluation to ensure that changes in the production process of one manufacturer will result in their connectors remaining compatible with connectors from another manufacturer.

It is easy to understand how a slight modification of the metals used in one connector (while being compatible with the metals in the mating connector by the same manufacturer) would not be compatible over long periods of time with the metals used in a connector by another manufacturer.

Similarly, a modification of the plastics used in one connector (while meeting the requirements to perform effectively with the plastics from the same manufacturer in the mating connector) may not be compatible over the long term with the plastics used by another manufacturer. Even minute differences in the thermal expansion rating of different plastics could pose problems.

These connector assemblies must remain electrically and mechanically secure for the very long power production life of PV modules. They must be watertight, not allowing water to reach the electrical connections. They must be mechanically strong and not separate under high wind loadings and ice loadings (and possibly thermal stress) to which they may be subjected. There can be no deterioration of the electrical contact due to a mismatch, however slight, of the metals used in each half of the mating connectors. When viewing the connector as a splice in the cable, the *Code* requires the splice to be at least as robust as the unspliced cable before the connector was inserted at that point [110.14(B)].

When connectors are mixed and matched, the listing and certification of the PV module as a complete assembly—including the connectors—becomes invalid. If a mixed, unmatched, pair of connectors fails either electrically or mechanically and causes damage at some time in the future, the listing of the module, and compliance with *NEC* requirements [including Sections 690.4(D) and 110.3(B)] could be a significant issue.

At this time, connectors should only be mated in pairs from the same manufacturer and in the same series. If equipment must be connected that uses connectors from different manufacturers, then some sort of cable connector adapter must be field manufactured or purchased that will allow the connectors to be mated in pairs from the same manufacturer. It is probably not a good idea to cut connectors off of a given product and change those connectors to a connector set from a manufacturer that will mate with the connector set on the other product. Although engineers at Underwriters Laboratories have said this procedure does not violate the listing on the product, several PV module manufacturers have stated that cutting the connectors off their listed module or other equipment will invalidate the warranty on that product. Therefore, it might be more appropriate to make a short length of cable with different connectors on each end to keep the mating pairs from the same manufacturer together, even though this will double the number of connectors in the circuit and possibly reduce the reliability of the system.

Photo 3.9 • Y cable adapter with MC-4 type connectors. All connectors must mate with matching connectors from the same manufacturer. Uncontrolled backfeed currents may pose problems.

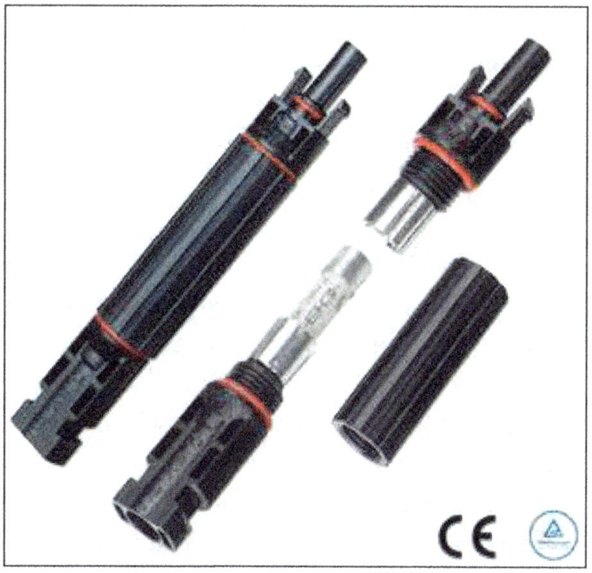

Photo 3.10 • Inline fuse with MC-4 type cable connectors. Connectors must be mated with cable connectors by the same manufacturer. Temperatures in installations exposed to sunlight may result in fuses being operated outside their listed rating. *Courtesy of Leader Group.*

A Warning. It is becoming more common to have to strings of modules combined with a Y-type cable adapter which may or may not have fuses inside the adapter or connected to the adapter for each input string (photo 3.9). Without fuses, the issue is: Are there any potential external currents that can damage the modules or string wiring? External currents may originate from additional combinations of module strings elsewhere in this circuit or from backfeed currents from the inverter. These are the sources of external overcurrents that must be addressed before this type of unfused paralleling can take place (690.9 Exception). In many cases, it is difficult to obtain information on whether the inverter can backfeed currents into faults in dc PV array wiring. These adapters are made by numerous manufacturers and, of course, the manufacturer of the Y adapter must be the same manufacturer who makes the connectors used on the PV modules and the connector used on the output cable from the Y adapter to ensure the listings on all connectors and modules remains intact. All MC-4 type connectors are not equal.

In situations where the Y adapter has fuses for both inputs internal to the adapter, or where inline cable fuses are used on each input, the environment that the fuses operate in must be considered (photo 3.10). Fuses listed for PV applications are listed with a maximum operating temperature of 50°C. On a typical rooftop application, it is probable that the adapter and fuses will be in an ambient temperature (around the cable/adapter) environment of 40°C to 50°C and be subject to solar heating from direct exposure to sunlight. Although the 2017 *NEC* has deleted any sunlight adders on temperature for cables that are more than 7/8 inch above the surface [310.15(B)(3)(c)], it seems evident that conductors anywhere within that 2014 *NEC* space of 36 inches will be subjected to additional solar heating. In a recent PV installation, my tools on an open stand about 12 inches above the roof in the sun became far too hot to handle without gloves in just a few minutes. It is likely that the fuses in these Y adapters will frequently be subjected to more than 50°C, even when they are behind the PV modules on a rooftop installation. This will mean the fuses are operating outside of the listed environment and the listing is invalid, which is a violation of the *Code*. I urge caution any time Y adapters, with or without fuses, are being used in a PV application.

Factory Training and Factory Tools Required

When it comes to installing connectors on cables in the field, caution is advised. In general, recognized components may not be installed to other equipment (or to a system by themselves) in the field, but are supposed to be installed only in the factory under factory-controlled conditions. This is a gray area in the UL Standards and in the *Code*. It is acknowledged, but not in writing anywhere, that a recognized component PV connector may be installed by a person having manufacturer's training on the installation of the connector and using manufacturer-approved tools to assemble the connector to the cable (photo 3.11). Because there are a significant number of different cables being used, the compatibility of a connector with a particular cable must be determined.

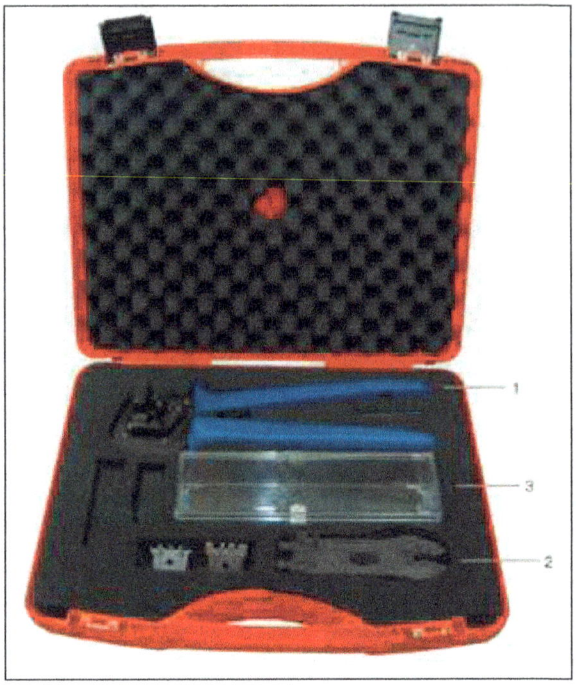

Photo 3.11 • Connector assembly tools. *Courtesy of MultiContact/STAUBLI.*

Module and Exposed Cables

Construction

All PV module manufacturers are using "PV cable/PV wire" fastened to their modules. (See 690.31.) Photovoltaic cable or PV wire is that cable meeting UL Standard 4703 for the use on modules and in exposed, field-installed PV source circuits. Of course, the old standby USE-2 conductors can also be used for exposed, source circuit wiring.

UL Standard 4703 allows a wide variety of materials and construction configurations to be used for the manufacture of PV cables so that they may be used both in the United States and in Europe. In Europe, the requirement for PV cables means that the cables must be fine-stranded, flexible cables and have tinned conductors. Photovoltaic system integrators and PV installers should exercise caution when purchasing bulk spools of PV cable or wire in the United States that may have fine stranding. It is difficult to find terminals on equipment, such as dc combiners and inverters, that can accept fine-stranded cables in the United States. Note the warnings on this problem in 110.14(A) and 690.31(H).

Although all PV cable or wire is marked sunlight resistant and has been tested for UV resistance with 720 hours of acceleration UV testing, some care must be exercised in selecting conductors that will withstand decades of use and exposure in PV environments. It is acknowledged that conductors with black insulation will last significantly longer than cables with colored insulation, and this applies to PV cable and wire and to USE-2. Also, conductors with synthetic rubber insulation (thermoset), such as XHHW-2, may last longer than conductors with PVC insulation (thermoplastic) such as THWN-2—although both can be used in the construction of PV cable/wire.

Photo 3.12 • Plastic wire ties, especially white ones (not UV rated), are likely to fail in a few short years.

Photo 3.14 • Many potential issues: EMT clamp and tape used to secure bundle of cables, plastic wire ties, foam sealant, and colored (non-black) PV wire insulation.

Photo 3.13 • Open conduit entry to dc combiner should not be used.

Photo 3.15 • Proper termination fitting for exposed conductors entering a dc combiner.

Mechanical Attachment

Even the highest quality cables and properly mated connectors will not survive decades of outdoor abuse, unless they are properly secured to the module and racking structure. Unfortunately, nylon cable ties—even the black ones marked UV resistant—are not withstanding the test of time in the PV environment. Again, these cable ties have not been tested with more than 720 hours of accelerated UV tests, which is equivalent to about two and a half years of outdoor exposure. In PV applications, particularly in the hotter Southwest regions of the United States, some of these are failing in just a few years (photo 3.12). Several manufacturers have brought to the market stainless-steel cable clips that are suitable for this application and will withstand the test of time. These clips or something similar, like the rubber insulated stainless-steel clamp shown in photo 3.6, should be used.

In a related area, the entry of these exposed conductors into conduits or dc combiners should be made with an appropriate conduit or opening termination fitting. Open conduits or conduits closed with nonelectrical foam products should not be used. (See photos 3.13, 3.14, and 3.15.)

The Conductors

As noted previously, single-conductor, exposed cables (type USE-2 or PV cable/PV wire) will be used for the module interconnecting cables. Both cable types are available in black (the most durable) and in multiple colors, including white and green (for exposed grounding and equipment-grounding conductors). As noted in 200.6(A)(6), black cable, even when smaller than 4 AWG, may be marked white as a grounded conductor, if necessary, at the time of installation where used as an exposed outdoor conductor at the array.

Normally, the exposed single-conductor cables are transitioned to a conduit wiring method when the circuits leave the PV array. Conductors in conduit, while they could be USE-2/RHW-2 (for flame and smoke retardant) or PV wire, are typically THHN/THWN-2 because they are less costly and the -2 rating is needed for the outdoor, wet environment and the high temperatures of conduit in sunlight [310.15(B)(3)]. Unfortunately, 14-10 AWG conductors with THHN/THWN-*2* insulation are not widely available due to low demand. Of course, THHN/THWN is available, but it does not have a wet, 90°C rating.

Demand will increase for the small-conductor THHN/THWN-2 conductors as inspectors start applying 310.15(B)(3) to rooftop HVAC installations. Due to the limited availability of 14-10 AWG THHN/THWN-2, XHHW-2 would be a suitable, superior alternative and may not be subject to additional deratings due to sunlight exposure or raceways more than 7/8 inch above the roof.

Although most PV arrays installed during the recent past have had the dc-negative conductor grounded (and colored white), newer arrays will be installed to meet the functional grounding requirements of the 2017 *NEC*. Of course, there are no designated color codes for ungrounded conductors, but common sense would indicate that ungrounded conductors would be clearly marked. It would be understandable if they were colored red for positive conductors and black for negative conductors installed in conduit.

However, many installations use black conductors for both and that is still acceptable under the *Code*. In positively grounded systems where the positive grounded conductor is colored white, the ungrounded negative conductor would be most clearly understood if it were black.

Functionally grounded PV systems would seem to require a "red is positive and black is negative" color coding (at least for conductors in conduit) because there will be no grounded conductor. However, both ungrounded conductors can still meet *Code* if they are black. As mentioned, basic black is preferred for all exposed outdoor conductors due to an increased UV resistance.

The newest bipolar PV arrays and bipolar inverters should not be ignored. In these systems, there will be red positive conductors, black negative conductors, and possibly white grounded conductors (where these conductors are solidly grounded). Of course, some installers will use black for both ungrounded conductors (acceptable under the *NEC*), and this will pose problems for the inspector and the troubleshooter when the cables are misconnected. If the "center tap" dc conductor is functionally grounded, then it, too, may be black. There are numerous circuit marking, routing, and grouping requirements in 690.7(C), 690.31, and 690.41.

As before, grounded conductors in PV dc disconnects should not be switched. Bolted, isolated, terminal-block connections are acceptable, but not required. Section 690.31(F)(1) will be clarified in the 2020 *NEC* and possibly by a TIA for the 2017 *NEC*.

Wiring Methods—Continued

All circuits in a PV system, as in other electrical systems, must be wired using a Chapter 3 or a 690.31 method that is suitable for the application and the environment. The circuits between dc PV disconnects and inverters are considered to be PV output circuits (and also inverter input circuits). They should be in a metal raceway or metallic cable assembly. Of course, local codes may dictate other requirements, such as the need to use raceways inside commercial structures for all electrical wiring.

PV Module Grounding

Grounding PV modules to reduce or eliminate shock and fire hazards is necessary but somewhat difficult. Copper conductors are typically used for electrical connections, and the module frames are often aluminum. It is well known that copper and aluminum do not mix (as was discovered from numerous fires in houses wired in the 1970s with aluminum wiring)—dissimilar metals are not to come into contact with each other (110.14).

Photovoltaic modules have aluminum frames. Many are anodized for color or have mill finishes and many are clear-coated. The mill finish aluminum (and any aluminum surface that is scratched) quickly oxidizes. This oxidation, and any clear-coat or anodizing, form an insulating surface that makes for difficult long-lasting, low-resistance electrical connections (in, for example, frame grounding). The oxidation and anodizing is not a good enough insulator to prevent electrical shocks, but it is enough to make good electrical connections difficult.

UL Standard 1703 (published by UL and developed and maintained by the Standards Technical Panel 1703) is used to certify and list all PV modules sold in the United States. It requires stringent mechanical and electrical connections between the various pieces of the module frame to ensure that these frame pieces remain mechanically and electrically connected over the life of the module. These low-resistance connections are required because a failure of the module insulating materials could allow the frame to become energized at up to 600 volts to 1500 volts (depending on the system design). The *NEC* requires that any exposed metal surface be grounded if it could be energized (250.4 and 250.110). Installers of PV systems are required to ground each module frame (690.43). *NEC* Section 110.3(B) and UL Standard 1703 require that the module frame be grounded at the point where a designated grounding provision has been made. The connection must be made with the hardware provided (if any) using the instructions supplied by the module manufacturer.

The designated point marked on the module must be used because this is the only point tested and evaluated by the certification or listing agency for use as a long-term grounding point. UL has established that using other points (such as the module structural mounting holes), coupled with typical field installation "techniques," *may not* result in low-resistance, durable connections to aluminum module frames. If each and every possible combination of nut, bolt, lock washer, and star washer could be evaluated for electrical properties and installation torque requirements, *and* if the installers would all use these components and install them according to the torque requirements, it might be possible to use the structural mounting holes for grounding.

New grounding devices are coming to market that will eventually ease the problem of module grounding. However, until they have been evaluated with specific modules and the module instructions address these devices, they do not meet the requirements of UL Standard 1703 or *NEC* Section 110.3(B).

The latest PV module grounding methods use engineered, manufactured mounting systems or racking systems that have been listed to UL Standard 2703 for PV module racks. This standard evaluates and certifies racking systems for mechanical strength, for electrical grounding continuity, or for both. A module is typically mounted with clips to the rack and, in many cases, the clips mount and ground the module to the rack so that only a single copper equipment-grounding conductor attached to the rack is needed to ground all the modules on that rack. The module instruction manual must show where and when top clips for mounting or grounding can be used. The listed rack instructions should indicate that the rack has been evaluated for mounting and grounding with a particular PV module. Efforts are underway to categorize PV module frames into categories related to size, material, strength, and coating. Then, rack manufacturers will have to test only a few categories to establish that racks will be able to mount and

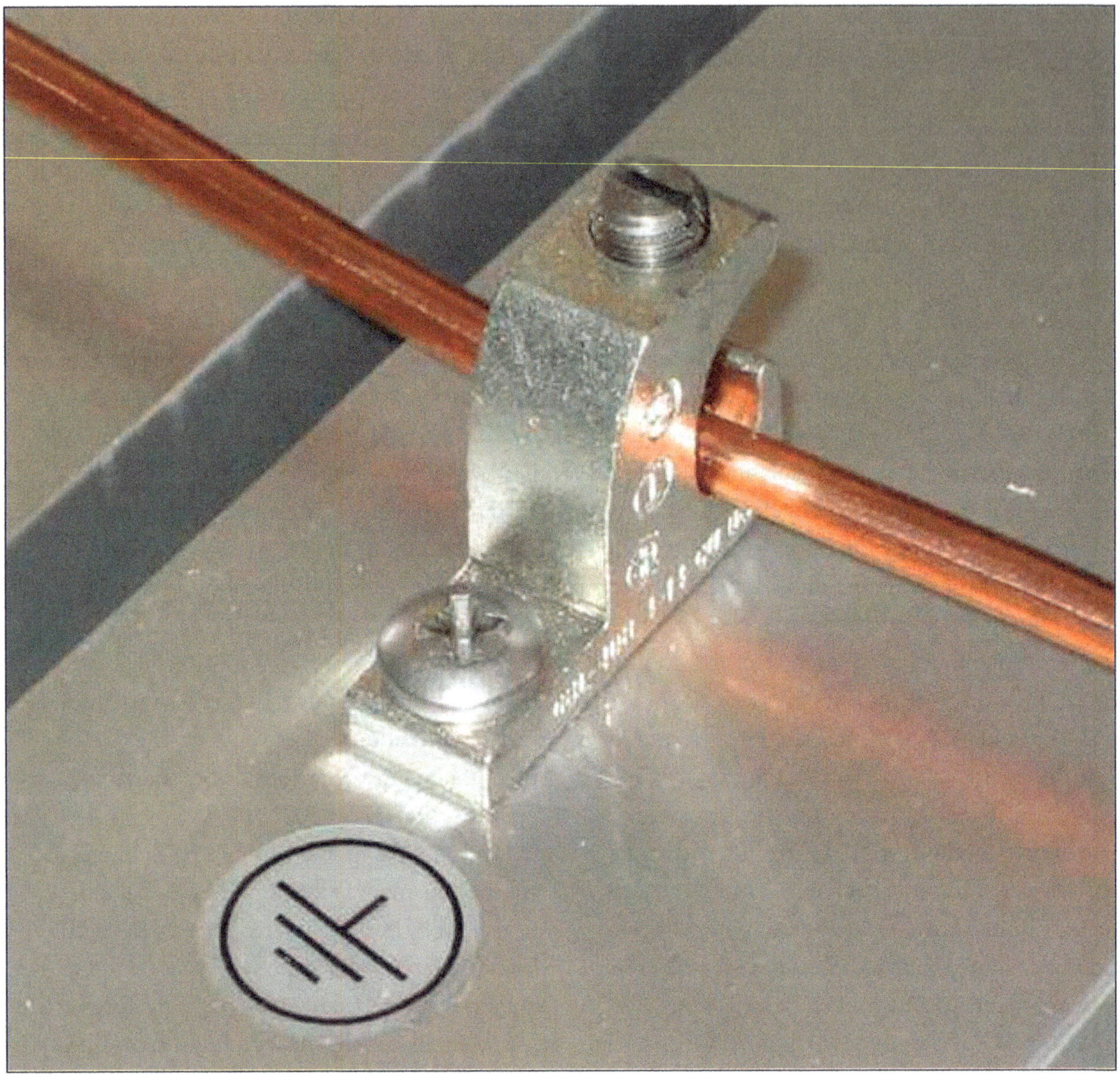

Photo 3.16 • Tin-plated copper direct burial lay-in lug.

ground numerous PV modules that are in those categories rather than testing each module.

Some PV module manufacturers are providing acceptable grounding hardware and instructions. Other manufacturers provide less-than-adequate (or no) hardware and unclear instructions. Revisions of UL 1703 are addressing these issues. In every case, the module manufacturer's hardware and instructions should be used (where possible) to ground the module at the points marked on the frame. UL Standard 1703 requires that the module manufacturer outline the specific grounding methods to be used, and either provide or specify the necessary hardware. These methods, and the hardware, will be evaluated during the listing of the module. It is likely that thread-cutting or thread-forming screws of the past will no longer be used.

For modules that have been supplied with inadequate or unusable hardware (or no hardware at all), there is one way to meet the intent of the *Code* and UL Standard 1703.

For situations requiring an equipment-grounding conductor larger than 10 AWG, a lay-in, tin-plated copper, direct-burial lug with a stainless-steel #10 screw, nut, flat washers,

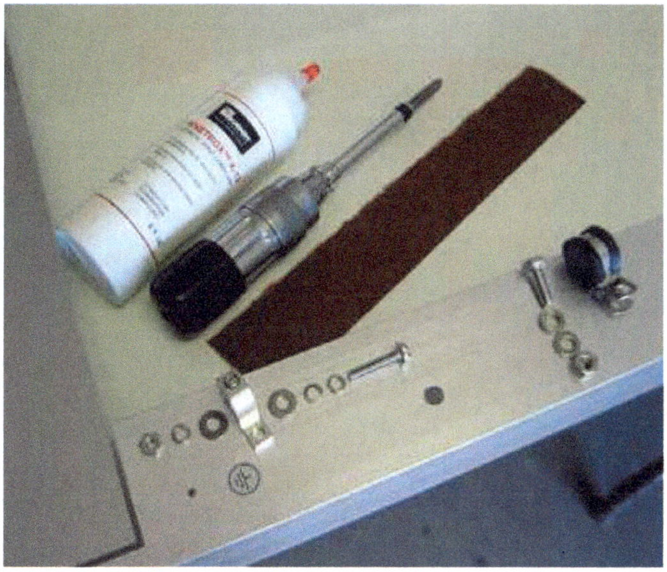

Photo 3.17 • Connecting tin-plated copper lay-in lug to aluminum.

Photo 3.18 • Grounding a metal roof—oops, outdoor rated lug and wire needed.

Belleville spring, and lock washers can be used to attach a direct burial lay-in lug to the module frame at the point marked for grounding (photo 3.16). Before attaching the lug to the module, a stainless-steel brush should be used to remove any anodizing, oxidation, or clear-coating from the aluminum module frame. A thin coat of antioxidant film should be placed on the clean aluminum surface. Flat washers are required to prevent lock washers from digging into the soft copper of the tin-plated lug or the aluminum of the module frame. The Belleville washer provides uniform tension, and a torque screwdriver should be used for all electrical connections (photo 3.17). Some new grounding lugs have been listed for use without the antioxidant compound because the design of the lug penetrates the oxidation. However, these should be evaluated with a specific module because of the varying thickness of the anodizing and clear-coat on the modules. It is not acceptable to use the hex-head, green grounding screws (even when they a have 10-32 threads) because they are not suitable for outdoor exposure and will eventually corrode. The same can be said for other screws, lugs, and terminals that are not suitable for outdoor applications.

The direct-burial lay-lugs are tin-plated lugs, made of solid copper, with a stainless-steel screw. They accept a 4 AWG to 14 AWG copper conductor. They are listed for direct-burial use and outdoor use and can be attached to aluminum structures (the tin plate allows this). The much cheaper tin-plated aluminum lay-in lugs look identical, but have a plated screw. They are not listed for outdoor use. If module grounding is to be done with a 14 AWG to 10 AWG conductor, then the lay-in lug may not be needed.

Other application-specific grounding lugs are available that will penetrate the module-protecting and -insulating coatings without the need for surface preparation of the module.

What size conductor should be used? The minimum *Code* requirement is for the equipment grounding conductor for PV source and output circuits to be sized to carry 1.25 times the short-circuit currents at that point, using the requirements of 250.122 and based on the overcurrent device

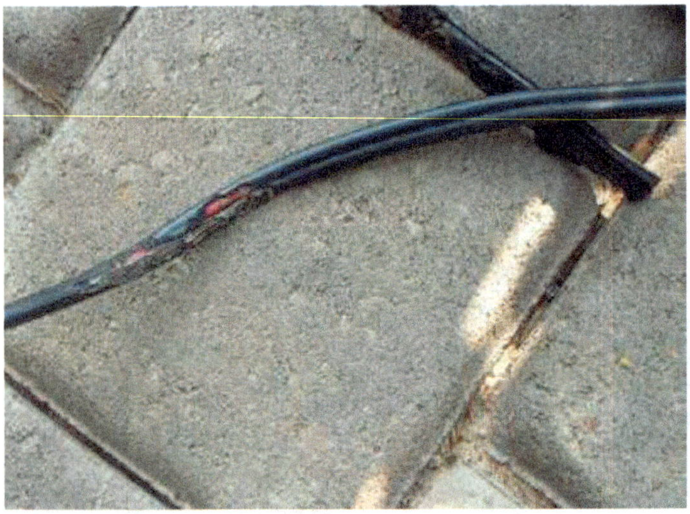

Photo 3.19 • Rodent-damaged conductors.

Photo 3.20 • Improper conduit terminations will allow moisture and rodents to enter the enclosure.

protecting these circuits. For PV source circuits where there are no overcurrent devices, an assumed overcurrent device per 690.9(B) shall be used per 250.122. While this may allow a 14 AWG conductor between modules, a conductor this small would require physical protection between grounding points. An equipment-grounding conductor smaller than 6 AWG can be routed behind the modules, from grounding point to grounding point, if the conductors are well protected from damage, as they would be when run in the module or rack frames in a roof-mounted array [250.120(C)]. If needed, an 8 AWG or 6 AWG conductor may be required (to meet the *Code* or to satisfy an inspector) and then lay-in lugs or a listed grounding device should be used (690.45 and 690.46).

The *Code* allows metal structures to be used for grounding and even allows the paint or other covering to be scraped away to ensure a good electrical contact. Numerous types of electrical equipment have parts that are grounded with sheet metal screws and star washers. This works (particularly in the factory environment) on common metals like steel, but not on aluminum due to rapid oxidation.

The racks and any metal roofing panels under the PV array should be connected to earth with equipment-grounding conductors where there is any possibility that these metal surfaces may be energized by conductor insulation failures (photo 3.18). Rodent damage and abrasion of conductor insulation could cause module frames, racks, or metal roofs to be energized (photo 3.19).

Single-conductor exposed wiring (USE-2 or PV wire) is allowed only in the near vicinity of the PV array to interconnect the modules and to return the end of the string conductor to the origination point of the string wiring. At this point, exposed wiring must transition to one of the more common wiring methods found in Chapter 3 of the *Code*, unless the conductors have a dual marking, such as RHW-2 or XHHW-2. Typically, this will be some form of raceway, such as electrical metal tubing. If the array output conductors penetrate the surface of the structure before reaching the first readily accessible dc PV disconnecting means, then they must be in a metal raceway where routed inside the structure. Metal raceways include rigid metal conduits, electrical metal tubing, intermediate metal conduit, metal wireways, and flexible metal conduit, and include Type MC metallic cable assemblies. The transition fitting keeps water, dirt, rodents, and other material out of the conduit. A rain head or a cord grip might be used (photo 3.20).

Photo 3.21 • Equipment-grounding conductor improperly connected to the module. (Incorrect hardware and incorrect installation procedure.)

UL Standard 1703—Grounding versus Bonding

The current edition (2012) of UL 1703, *Standard for Flat-Plate Photovoltaic Modules and Panels*, delineates the differences between *grounding* and *bonding*. Bonding refers to the factory-made electrical connections between the four or more aluminum sections of the module frame. Grounding refers to the field-installed electrical connection between the aluminum module frame and the equipment-grounding system (usually consisting of copper conductors).

Bonding the frame pieces together at the factory, using very specific materials and methods, results in a durable electrical connection between the frame pieces (so that any failure in module insulation or external conductor insulation will result in all pieces of the frame receiving equal voltage). Factory bonding also ensures that when the module frame is properly field-grounded at one of the marked and tested points, the entire module frame is maintained at the ground (earth) potential under fault conditions.

During the bonding process, all screw fasteners (when used) are precisely torqued to the specified value by automated equipment or by trained technicians using torque screwdrivers. Factory bonding materials and methods are evaluated for low resistance and durability during the certification and listing processes. After the listing, if the manufacturer changes any of the bonding materials or methods, the changes must be reevaluated by the listing agency. The materials (including any screws or washers) are not specified generically—they are specified to the original equipment manufacturers and must always be obtained and used from those sources unless any change is reevaluated by the certification or listing agency.

Compare this precisely controlled and evaluated factory bonding system with field-installed grounding techniques used to connect copper equipment-grounding conductors to aluminum module frames. Grounding a PV module is difficult at best, for a number of reasons. The first is that module manufacturers may not realize the importance of this connection to the overall safety of the system. The revised UL 1703 distinguishes the differences between bonding and grounding. Manufacturers may have the impression that bonding techniques and materials used in the factory may be applied to grounding connections made by installers in the field. Instruction manuals and hardware (sometimes supplied) show techniques that are not always consistent with good electrical connections (photo 3.21). Field-made connections using threaded fasteners are rarely torqued to the specified value, even when that value is given in a module's instruction manual (because not all PV installers have or carry torque screwdrivers or wrenches). The *NEC* requires that where threaded fittings or connections are provided with a torque value, a calibrated torque device shall be used to tighten that connection or terminal [110.14(D)]. Field-grounding connections may or may not be inspected by AHJs, and they are never tested for overall continuity. Also, because PV systems can operate without trouble for many years, there may be little motivation to inspect these connections after the original installation.

04

The Inverter—Operation and Connections

Utility-interactive inverters range in size from 175 watts (photo 4.1) to 2.5 megawatts and come in all shapes, sizes, and colors (photos 4.2 and 4.3). New models are being introduced monthly. These inverters will be listed by UL, CSA, ETL, and TUV Rheinland. All of these organizations are designated as nationally recognized testing laboratories by OSHA for testing and listing PV modules, inverters, combiners, and charge controllers using standards published by UL.

Some inverters have only a single set of dc input terminals and no internal dc or ac disconnects. With these designs, an external dc PV disconnect must be installed (690.15). Even if the inverter has more than one set of input terminals for paralleling separate strings (source circuits) of modules, external dc PV disconnects must be used on each input (photo 4.4).

Photo 4.1 • Westinghouse/Andalay AC PV modules using Enphase microinverters.

Other inverters have internal dc disconnects or disconnect housings that attach to the main inverter section containing the electronics package. The method used to mount the internal disconnects, the ease and accessibility of the disconnects, and the way they are separated from the inverter proper vary from brand to brand and from product to product. Installers and AHJs must reach a mutual conclusion on the suitability of these disconnects for meeting the various isolation and disconnect requirements in the *Code*.

Because the inverters are listed with the disconnects, it can be presumed that the disconnects are properly rated for the dc load break operation and that they would meet the requirement for the isolator and disconnect requirements of 690.15.

Meeting the requirement for equipment isolators and disconnects (690.15) will

Photo 4.2 • KACO utility-scale inverters.

Photo 4.3 • SMA Sunny Central 2.5 MW utility-scale inverter. *Courtesy of SMA-AG.*

Photo 4.4 • Inverter with internal dc disconnects plus additional external dc and ac disconnects.

require additional considerations. If an inverter required factory service, could the energized PV source or output circuits be disconnected from the inverter safely when there is no external disconnect? If a disconnect housing is attached to the inverter and that housing does not have to be removed to service the inverter, then some degree of safety is ensured. However, if energized conductors must be disconnected from internal switches and pulled through small conduit knockouts, the situation must be examined carefully. Will qualified people, who know how

Photo 4.5 • Transformerless inverter with internal dc disconnect.

Photo 4.6 • Inverters with attached dc disconnects that can be separated from inverters.

to disable the array, be doing the removal? Or will an unqualified person try to pull energized conductors through the knockouts? (See photos 4.5 and 4.6.)

Peak Power Tracking

A PV array is a current source of energy and the output power depends on the load that the inverter places on the array. No loading (zero current) would operate the array at the open-circuit voltage point (V_{OC}). The heaviest loading (a short-circuit—hard to achieve) would operate the array at the short-circuit current point (I_{SC}). Neither of these operating points would produce any power output from the array, power being the product of voltage and current and (at these points) one of them is zero. However, for every condition of sunlight intensity (irradiance) and array temperature, there is a load that will extract the maximum power from the array that the array can produce under those conditions. The utility-interactive inverter will find that maximum or peak power point—maximum power voltage (V_{mpp}) and maximum power current (I_{mpp})—and track that point as the sunlight and temperature vary throughout the day. See chapter 2 for more information.

Automatic Operation

Today's utility-interactive inverter is designed, manufactured, tested, and certified or listed to operate automatically in the PV system. There are no transfer switches. The inverter seamlessly converts dc power from the PV array into ac power that is fed to the utility-supplied premises wiring system. The output of the inverter is functionally connected in parallel with the premises' wiring (and loads) and the utility service.

One of the most important aspects of the inverter is the anti-islanding circuit. An anti-islanding circuit is designed to keep the utility electrical system (both premises wiring and utility feeder) safe in the event that the utility is being serviced or is disconnected at some point in the transmission system, distribution system, or premises wiring system.

Unlike an engine-driven generator, which can feed power into a blacked out or disconnected local utility feeder system, an inverter anti-islanding system prevents an inverter from energizing the "dead" power system.

This circuit prevents the inverter from delivering ac power if the utility voltage and frequency are not present, or if they are not within narrowly defined limits. This circuit monitors the voltage and frequency at the output terminals of the inverter. If the voltage varies more than plus ten percent or minus twelve percent from the nominal output voltage the inverter is designed for (120, 240, 208, 277, or 480 volts), the inverter cannot send power to the output terminals. Nor is there any voltage on these terminals (from the inverter) when the inverter shuts down if the circuit from the utility is at zero voltage. In a similar manner, if the frequency varies from 60 Hz to more than 60.5 Hz, or less than 59.3 Hz, the inverter also cannot send power to the ac output. If the utility power is suddenly not present at the output terminals for any reason (inverter ac output disconnect opened, service disconnect opened, meter removed from the socket, utility maintenance, or utility blackout), the inverter senses this and immediately ceases to send power to the output terminals.

The anti-islanding circuit in the inverter continues to monitor the ac output terminals and when the voltage and frequency from the utility return to specifications for a period of five minutes, the inverter is again able to send PV power to the ac output. When the inverter is not processing dc PV power into ac output power, it essentially stops taking power from the PV array by moving the load on the PV array to a point where there is no power. Usually this is the open-circuit voltage point for the PV array.

Circuit Sizing

Direct Current

The dc input circuit and conductors to the inverter are sized based on the dc short-circuit current in those conductors (that sizing is covered in previous chapters). There is no direct relationship, or *Code* requirement, between the array-rated output current and the manufacturer's specified maximum dc inverter input current. Normally, a PV array is rated in watts at standard test conditions of 1000 watts per square meter (W/m^2) of irradiance and a cell temperature of 25 °C. In most cases, an array will operate, on average, at a lower power output because of normal and expected power lost due to module heating. For this reason, inverter manufacturers typically suggest sizing PV arrays (standard test condition dc watts) at ten to twenty percent greater than an inverter's ac power output rating. It does no short-term harm to connect an even larger PV array to an inverter because the inverter must limit its output to the rated value no matter how much array power is applied. If this oversized array is used, the inverter will spend more operating time each day at rated power output than it would with a smaller array. The penalties for designing a system in this manner will be increased module cost for the larger array; some lost power on sunny, cool days; and possibly a slight reduction in the inverter's lifespan due to longer operation at full-power higher internal temperatures. However, systems installed where microclimates reduce irradiance during portions of the day, or where the local utility limits the inverter-rated power output, this type of operation may result in more energy being delivered from the output than would be possible with a conventionally sized system.

DC Input Fusing

Some models of small (<10 kW) and large (>100 kW) inverters have dc input fuses mounted inside the inverter or inside a combiner or disconnect device attached to the inverter. The smaller fuses (30 amps or less) are usually mounted in "finger-safe" fuse holders that allow the fuse to be safely replaced in an unenergized (no load current) state.

However, when fuse ratings go over 30 amps, with values as high as 400 amps or more, these

Photo 4.7 • PV combiners with internal disconnects on the outputs of both polarity conductors.

Photo 4.8 • Disconnects, with fuses, for each input installed near the inverter

fuses are mounted in exposed fuse holders or bolted directly to a dc busbar. One side of each fuse is tied together with the dc input of the inverter. The other side of each fuse is hardwired to the output of a PV dc combiner and these combiners will be scattered throughout the PV array—sometimes over thousands of acres of real estate. Although the inverter can be turned off and the dc input capacitors allowed to discharge (up to five minutes), each fuse is still energized from its own input and the combined inputs of all the other fuses through the common busbar. The only way to safely service these fuses is to go through the entire PV array, find all of the combiners, and open or pull each and every source circuit fuse (those less than 30 amp "finger safe" fuse holders). An optional disconnect at the output of every combiner speeds this process and makes servicing the combiner fuses safer, but all disconnects must be located and opened (photo 4.7). Combiners are available with internal load-break-rated output disconnects that, if used with a remote control, could speed the shutdown and possibly meet some of the rapid shutdown requirements (690.12).

When these fuses are present in the input of larger inverters, the safest way to provide service is to install a dc disconnect near the inverter on each dc input to a fuse (photo 4.8). These grouped disconnects can be opened easily, and with the inverter turned off, the fuses can be safely removed in a de-energized state. Section 690.15(C) and 240.40 require dc disconnects near these fuses so that they can be safely de-energized from all sources when being serviced.

Alternating Current (ac)

The ac output circuit of an inverter must be sized at 125 percent of the rated output current of the inverter (690.8). Some inverter manufacturers specify the rated current or a range of values (due to varying line voltages from nominal). If this specification is not given, then the rated power may be divided by the nominal line voltage to determine a rated current. For example, a 3300-watt inverter operating at a nominal voltage of 240 volts would have a rated current of:

3300 watts / 240 volts = 17.2 amps
1.25 x 17.2 = 21.5 amps

This would be the circuit ampacity requirement and it would usually be protected by a 25-amp overcurrent device at the utility end of the circuit.

These inverters are not capable of providing any sustained (more than a second) surge currents, so the rated output current is all that can be delivered. When faced with a short-circuit, the rated output current is all that can be delivered, but—more than likely—the reduced line voltage due to the fault will cause the inverter to shut down. With the available maximum current limited to the rated inverter output current and the overcurrent device rated at 125 percent of that current, it is doubtful that the overcurrent device would trip or blow from currents from the inverter. That device would, of course, activate on the high available fault currents from the utility.

Connections

Dedicated Circuit

NEC 705.12(B)(1) requires that the inverter output be connected to the utility power source at a dedicated disconnect and overcurrent protective device (OCPD). In most systems, this is a backfed breaker in a load center/panel board [705.12(B)]. Inverters may not have their outputs connected directly to another inverter or directly to an ac utility-supplied circuit without first being connected to the dedicated disconnect or OCPD. Utility-interactive microinverters and ac PV modules are exceptions to this rule because they are tested and listed to have multiple inverters connected in parallel on a single circuit with only one OCPD or disconnect device for an entire set of inverters. More details on the requirements found in 705.12(B) of the *NEC* and earlier editions will be found below.

The OCPD must be sized at a minimum of 125 percent of the rated inverter output current (or the total of the output-rated output current from multiple microinverters or ac PV modules). It must also protect the circuit conductor from overcurrents from the utility side of the connection. It is usually *not* a good idea to install a larger OCPD than the minimum required value (but allowing a round up to the next standard value is okay and needed) because the inverter may, as part of the listing or instructions, be using the OCPD to protect internal circuits. The maximum value of allowable overcurrent device should be in the inverter's instruction manual and should not be exceeded [110.3(B)].

Where fused or unfused disconnects are used for the ac inverter output disconnect (or the required utility disconnect), the circuit connected to the utility source should be connected to the supply side (top) of the disconnect with the inverter ac output connected to the load side (bottom). A warning label is required, because even when the disconnect is opened, the inverter ceases to produce power within a fraction of a second and the exposed load-side terminals pose no shock hazard. However, the connected utility side terminals, while protected, are still energized.

GFCIs and AFCIs

The ac output of a utility-interactive inverter should *not* be connected to a ground-fault circuit interrupter (GFCI) or to an arc-fault circuit interrupter (AFCI) circuit breaker, as these devices are not currently designed to be backfed and will be damaged if backfed. These devices have terminals marked "line" and "load" and have not been identified, tested, or listed for backfeeding.

Ground-Fault Protection on Main Breakers

The *NEC* has had, for many editions, a requirement (230.95) that solidly grounded wye services rated at 150–600 volts phase-to-phase and 1000 amps or more have ground-fault protection (GFP). Connecting a PV inverter ac output on the load side of these GFP-equipped main breakers may pose safety issues.

Circuit breakers are manufactured with numerous optional accessories, including (depending on the manufacturer and model) shunt trips, auxiliary switches, remote indicators, power operation, adjustable trips, and GFP trip mechanisms. While UL Standard 489 requires tests for evaluating the backfeed suitability of the basic circuit breaker, most of the accessories are not evaluated for backfeeding. In fact, backfeeding may have no

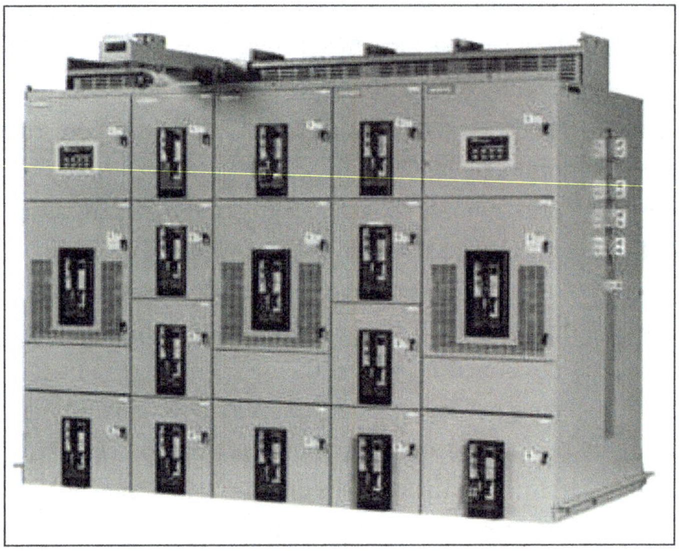

Photo 4.9 • 1200-amp, 480-volt ground-fault-protected switchgear. To backfeed or not? Only the manufacturer knows.

effect on many of these accessories, and specific testing for backfeeding may be unnecessary.

However, older and possibly some current ground-fault trip mechanisms may be damaged if the circuit breaker has voltages on both line and load terminals after the breaker has been opened by a ground-fault trip. UL 489 testing does not evaluate backfed GFP main breakers in a manner that subjects the ground-fault device to the conditions it would experience in a utility-interactive PV system or possibly even in a parallel-connected generator installation where line and load terminals are both energized during and after a ground-fault trip. Photovoltaic inverters may have energized outputs up to two seconds after the ac utility power is removed from the output.

Circuit breaker manufacturers should be evaluating all accessories supplied with their breakers for operation under all possible application configurations. However, utility-interactive inverter installations are relatively new applications and inverters are being installed in electrical installations that may be decades old.

Informal discussions with manufacturers indicate that most new designs use a current transformer to power a GFP device, and that current transformer does not respond to voltages on a tripped open main breaker. The GFP device, powered by the current transformer, should not be damaged by backfeeding.

However, there is some confusion and uncertainty about the older GFP/breaker designs that are installed widely and may still be on the market (photo 4.9).

Although 705.32 and the Exception no longer require that ground-fault protection devices be identified and listed for backfeeding, 110.3(B) requires that they be used according to their listing and labeling. Ground-fault protection devices attached to circuit breakers are not specifically tested and listed under UL Standard 489 requirements.

Only the breaker manufacturer (the design engineer) can verify that the particular breaker with GFP (by part number and model number) will not be damaged under PV inverter backfed conditions, where both line and load terminals are energized at the same time during and after the GFP device trips the breaker. Bulletins and information from sales departments are usually insufficient to make a determination.

After determining that the main breaker/GFP is suitable, then the issue of protecting the load circuits under ground-fault conditions from all sources (utility and PV) must be addressed (705.32 EX). An engineering analysis would be required that shows how and where ground-fault currents are sourced. What are the impedances involved in the utility source and the PV source? How much current can each supply under varying types of ground faults? Ground faults are not always "hard" low-resistance faults and may be arcing faults of varying impedances. Suppose the PV system sources enough ground-fault current to prevent the main breaker GFP from tripping? How is the ground fault contained or interrupted?

Some manufacturers are wary of putting some sort of GFP device on the inverter output because this is a nonstandard connection and any ground faults detected might only be those originating between the device and the inverter, not load ground faults.

Some manufacturers have a main breaker GFP that can take inputs from multiple ground-fault sources like dual utility feeder systems. But these would be found in limited, special instances where there are multiple utility feeds.

Summary

The requirements of the *NEC* are stringent, but can be met. There are no one-size-fits-all solutions to this issue. The following steps should be followed before connecting a PV system that could backfeed a GFP breaker. There may be others.

1. Accurately determine that all ground-fault protection devices are suitable for operation in a backfed manner with a utility-interactive PV inverter.
2. Select an appropriate GFP device that can be connected to the inverter's outputs to control ground-fault currents from that source.
3. Make an engineering assessment of the magnitudes of the potential and available fault currents from both utility and PV sources to the load circuits being protected. Circuit impedance calculations under fault current levels for all sources and the load impedance should be made.
4. Determine the proper setting for all adjustable-trip ground-fault protection devices that will ensure that the load circuits are protected from all ground-fault current sources.
5. With the GFP breaker being back fed with current from the PV inverter, the GFP device should be tested using the internal test device. It should be tested a second time to ensure that the device was not damaged during the first test.

In many cases, it may be easier to implement a supply-side (of the main GFP breaker) PV connection as allowed by 705.12(A) and 230.2(A)(5) / 230.82(6). These connections will be discussed in a later chapter.

Functional Grounded Systems

It is likely that the type of PV system being installed (with respect to grounding) will be determined by the inverter topology. That topology will, in the future, most likely result in a functional grounded PV system. The internal methods by which the inverter achieves the functional grounding may, to some extent, determine the dc input and ac output circuit configurations. The inverter instruction manual will have the appropriate instructions.

Newer Inverter Technologies

Microinverters

The inverters that have been covered above are known as "string inverters" because they operate with a string of series-connected PV modules. These inverters range in power from more than one megawatt down to approximately 700 watts. DC maximum system voltages range from 125 volts to 1000 volts and higher. Inverters connected to one (or possibly two) PV modules are classified as microinverters.

The Enphase microinverter is a typical example (photos 4.10 and 4.11). It is a small inverter (hence the name) that is designed to work with a single PV module and to operate at a maximum of approximately 70 volts on the dc input. The inverter is connected directly to the PV module using the existing conductors and connectors (now locking in most cases) attached to both the module and the inverter. Available units are rated in the 170-watt to 500-watt range, but as with

Photo 4.10 • Enphase 400 W microinverter. *Courtesy of ENPHASE.*

Photo 4.11 • Westinghouse AC PV module with Enphase inverter showing ac and dc cables.

other PV products, ratings and specifications change continually.

The microinverter is a utility-interactive inverter. As part of a PV system (module and inverter), it must have dc ground-fault protection [690.41] per the requirements of UL 1741, just as is required in larger inverters.

Microinverters have ac input and output cables and connectors and are listed in a manner that allows multiple inverters to be connected with up to about 15 units on a single output cable (called a trunk cable) (photo 4.12). The dc and ac connectors used with the Enphase microinverter have been listed as load-break-

Photo 4.12 • Multiple microinverters connect to a trunk cable. ANDALAY/WESTINGHOUSE ac PV modules with trunk cable integrated into module frames.

rated disconnects. With a power output in the 240-watt to 400-watt range (depending on the model), the rated ac output current at 240 volts will range from 1 amp to 1.2 amps. On the 14 AWG cable with a 15-amp overcurrent device, the rated continuous current for that circuit is limited to a maximum of 12 amps. This rating will allow 1 to 12 inverters to be installed on the same ac output cable. Trunk cables with larger conductors will allow greater numbers of micro-inverters to be connected.

AC Output Circuits

The output circuit of any utility-interactive inverter, including microinverters, up to the first overcurrent protection device (OCPD) is very much like an ac branch circuit. If the utility voltage is removed from this circuit (for any reason), the circuit becomes de-energized (dead)—just like a branch circuit. If there is a line-to-line or line-to-ground fault on this circuit, the OCPD responds in a normal manner to the fault currents generated by the utility. The inverter can generate no more than its rated current per UL Standard 1741, and when the fault occurs, the drop-in line voltage will normally cause the inverter to shut down. When the branch circuit breaker opens in response to the fault, the inverter shuts down.

It would appear that these inverter output circuits could be wired using any Chapter 3 wiring method suitable for the environment (indoors, or hot, wet, and UV outside, and hot in attics).

The metal raceway requirement of 690.31(G) applies only to the always-energized dc PV source and dc PV output circuits on or in a building.

Flush-mounted inverters are starting to appear on the market, and they will have a ventilation system to remove the internal heat generated during the inverting process.

An ac GFCI device should *not* be used to protect the dedicated circuit to the microinverter or ac PV module, even though it is an outside circuit. None of the small GFCI devices (5 mA to 30 mA) are designed for back feeding and will be damaged if backfed. In a similar manner, most ac AFCIs have not been evaluated for backfeeding and may be damaged if backfed with the output of a PV inverter.

Combing Multiple Sets of Microinverters or AC PV Modules

In multiple strings of inverters, there is no *NEC* requirement that an ac combining panel (load center) be located on the roof. In fact, most NEMA 3R load centers must be mounted against a surface to keep water from penetrating holes in the back panel. Such a surface may have to be added to properly mount a 3R load center on a roof. There might be problems meeting 110.26 clearance requirements.

A further issue with an OCPD on a roof is heating of the device over its rated 40°C operating temperature. Gray load centers in the sun will normally operate 10°C to 20°C hotter than the local ambient temperature (which, in some cases could be 40–50°C). This may be difficult to compensate for when considering available equipment, the size of ac conductors attached to inverters, and listing restrictions on the inverters. Nevertheless, it is possible to mount an ac load center on a roof with proper solar shielding and use it to combine the outputs of U-I inverters or sets of microinverters.

The rating of any combining panel and the ampacity of conductor from that panel to the backfed breaker in the main load center (as well as the rating of the main load center and the backfed breaker) must meet 705.12(B) requirements that are covered in Chapter 7.

The ac output conductor for a set of inverters must have an ampacity of 125 percent of the continuous currents for all the inverters on that circuit. The backfed circuit breaker in the panel must be rated the same, and if an odd current rating is determined, the breaker rating should be the next larger size. The breaker must protect the conductor under the conditions of use, and the conductor ampacity must be derated for those conditions of use.

The ac output circuit from each set of inverters must have an equipment-grounding conductor to facilitate OCPD operation during ac ground faults. Some microinverters have a three-wire output through a four-contact connector. The unused terminal in the connector is reserved for future use. The three active pins in the connector are 240 V L1 and L2 and a neutral. There is no ac equipment-grounding conductor. This lack of an equipment-grounding conductor in the cable requires that the equipment-grounding conductor for the microinverter or ac PV module be an external connection to the inverter case, where the case is metal. This external equipment-grounding conductor must be connected to the fixed wiring system (usually, but not always conduit) where that wiring system originates. The ac module or microinverter instructions will cover grounding requirements and should be followed [110.3(B)].

Unless the microinverter bracket has been designed and evaluated as a grounding or bonding jumper, grounding the microinverters does not ground the rack or the modules (and vice versa).

There is only one ac neutral-to-ground bond in

Photo 4.13 • SUNPOWER AC PV Module. *Courtesy of Sunpower.*

an ac electrical system. That bond is made in the existing service-entrance equipment. No additional neutral-to-ground bonds should be made when installing a PV system unless a supply-side service-entrance connection is made or transformer isolation establishes a separately derived system.

AC PV Modules

Take a normal dc PV module in the factory and connect a microinverter to it, fasten the microinverter to the back of the module, and you have an ac PV module (photo 4.13). AC PV modules are becoming more common in the market. One is the ac PV module by Westinghouse Solar/Andalay Solar and it has a unique frame that is also the module mounting rack (photo 4.12). Photo 4.14 shows an LG/Enphase ac PV module. Because the dc wiring between the module and inverter is listed with the entire ac PV module and becomes an integral part of the product, dc requirements in the *Code* no longer apply to the ac PV module. The ac PV module is a utility-interactive device and has a similar ac output cabling system to the microinverter addressed above.

DC Connections

In a standard PV module/microinverter combination, the microinverter dc connection to the PV module may have to be disconnected to replace the microinverter should it or the module fail. While the voltage will be a maximum of approximately 70 volts with current inverter designs, the current may be in the 3-amp to 7-amp range and the connectors could possibly be damaged at this voltage and current, posing a possible safety hazard. While very few inspectors may request a costly and impractical load-break-rated disconnect, the *Code*-compliant solution is quite simple. The back of the PV module must be accessed to reach these dc connections, and this generally requires that the module be unfastened from the mounting system. Because the module is accessible and *is* being accessed, just by putting a blanket or other opaque material over it per 690.18 (2104 *NEC*) will reduce the dc output voltage and current (and the ac current) to near zero, allowing the module/inverter dc connectors to be safely opened. Opening this connection with the module blacked out will likely be safer than opening the same connectors on a module in a high-voltage string of modules.

Each microinverter or ac PV module will have an ac output trunk cable to allow the multiple inverter parallel connections. This cable may carry currents in bright sunlight of 1 amp or more at 240 volts (in current designs) from the first module/inverter in the set to as much as 12 amps at 240 volts through the last connector of the set that has multiple series-connected devices. Servicing the single ac PV module or utility-interactive microinverter could be accomplished by covering the module to reduce the dc and, hence, the ac current to zero.

However, not covering all modules in the set would allow current from other, non-covered, modules/inverter to flow through the cable and

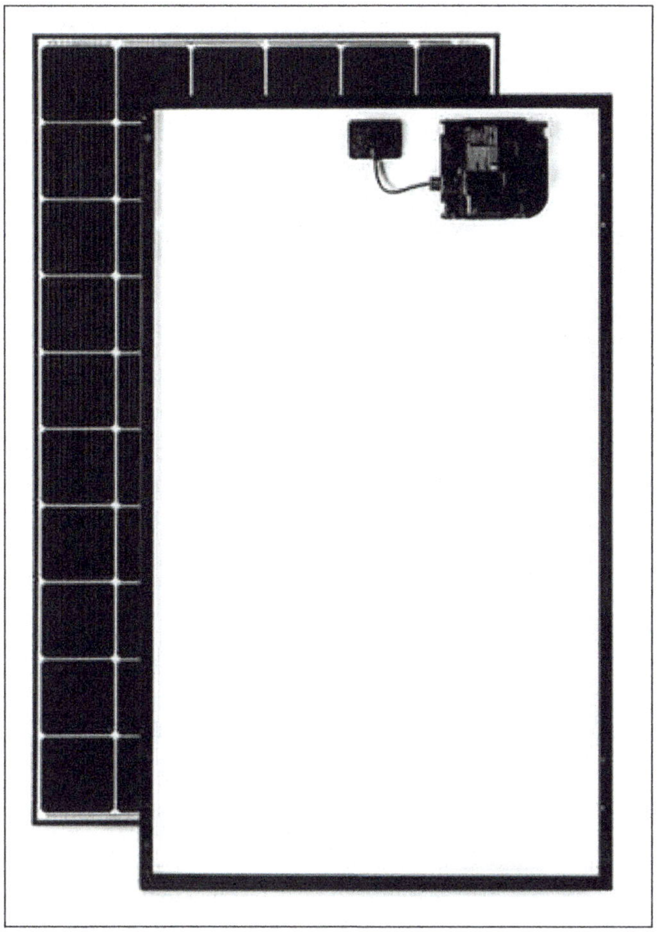

Photo 4.14 • AC PV module, LG NeON 2 AC, as defined in 690.2 and 690.6. *Courtesy of Enphase.*

at 240 volts, could damage the connector and possibly pose a shock hazard when opening these ac connections under load. To some extent, the hazard is minimized because the inverter anti-islanding circuits shut down very rapidly, reducing any arcing when the ac connector is opened.

Opening the ac circuit at the PV backfed breaker in the building service-entrance panel would be a safe solution if that breaker could be locked open. However, breaker locks are few and far between and lock-out/tag-out procedures are not generally used in residential and commercial electrical systems.

Where disconnect/isolation requirements for each microinverter can be met by the use of connectors on the dc input and ac output, the installation of a separate ac disconnect on the roof near the ac PV modules or microinverters will enhance safety, but is not a *Code* requirement. A common 60-amp unfused, pullout, air-conditioning disconnect costs less than $10 at building supply centers. It provides the disconnect; a place to terminate the ac trunk output cable from a single set of microinverters or ac PV modules; a place to originate the field-installed wiring system to the ac load center in the house; and is usually cheaper than a separate junction box and cover.

Some microinverters and ac PV modules may have dc and ac connectors fully listed as load-break-rated disconnects. The Enphase microinverter has both the ac and dc connectors listed as load-break-rated disconnects. Such a rating would meet 690.15 disconnecting/isolating means requirements and the separate rooftop disconnect would not be required (although the device mentioned above would still be an inexpensive transition point between the inverter/module ac cord and the conduit conductors leading from the roof).

AC PV Module Maintenance —A Gray Area

Combinations of a microinverter and a PV module with exposed dc connectors and dc conductors between the PV module and the microinverter are being certified and listed as ac PV modules. Some of these products have instruction manuals that say the microinverter may *not* be removed or disconnected for service from the PV module to affect a repair of either component. In fact, such a separation may be difficult in the field and would violate the listing on the ac PV module and violate the warranty.

Other manuals give specific instructions for removing the microinverter from a PV module for repair. At issue is the definition of an ac PV

Photo 4.15 • Numerous parts are required for a string-inverter system.

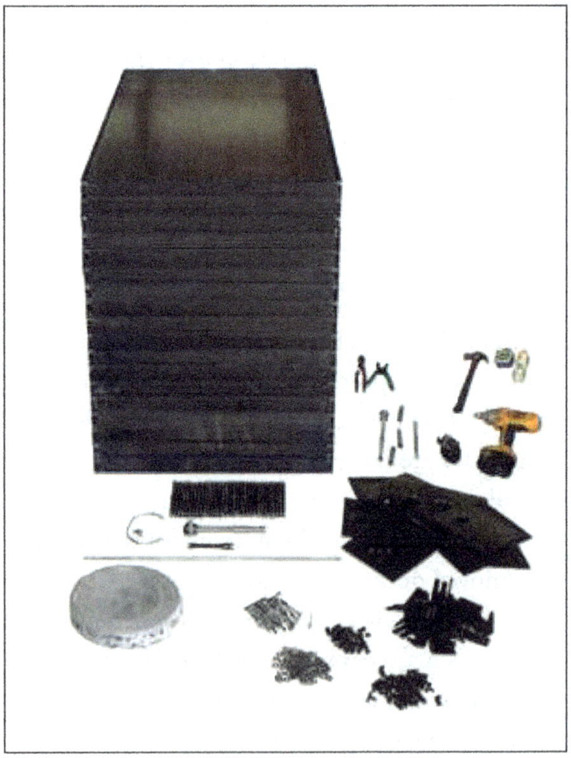

Photo 4.16 • Far fewer parts are needed to install the Andalay AC PV module system, which includes the rack built into the heavy module frame.

module as a factory assembled unit and the 690.6 exemption from all dc *Code* requirements for these products with exposed dc connectors and dc conductors. However, exposed connectors are subject to loosening or being opened in the field. Connectors and conductors are exposed to environmental degradation, ground faults, arc faults, and animal damage.

Also at issue is the microinverter-to-PV module frame bonding when the mechanical or electrical connection is broken in the field. When the microinverter is replaced, how is the bonding connection quality verified and how is the certification or listing maintained without NRTL evaluation?

At some point, these issues will be addressed in UL Standard 1741 and possibly in the *NEC*.

Advantages Will Boost the Market

The use of microinverters and ac PV modules will proliferate due to several advantages they offer over conventional string inverters.

The first is a simplified set of installation requirements and a reduced number of separate parts. See photos 4.15 and 4.16 for quantitative differences in the amount and types of equipment involved in installing an ac PV module system versus a conventional string inverter system.

In a dc series-connected string of PV modules, module mismatch is sometimes an issue that affects the string performance. Modules come out of the factory with slight (up to 10%) variations in performance. The string of modules in a dc system cannot deliver current above the current delivered by the weakest module in the string. The mismatch between module currents results in some lost power compared with a dc string of modules that is equal in every specification.

Photovoltaic modules near the top of an array on a sloped roof may operate hotter than modules lower down on the roof due to hot air rising behind the modules. Depending on how each string of modules is connected, some loss of power may

Photo 4.17 • DC-to-dc power converter attached to the back of each PV module. *Courtesy of SoarEdge.*

occur if hot modules are connected in series with cooler modules.

Shading is also a problem in a conventional string-inverter configuration. The shading of a single module will result in a power loss from that module, but may also reduce power from the other, non-shaded modules in the string.

The microinverter and the ac PV module work at the individual module level. Each inverter extracts the maximum power from that module no matter what the other modules in the PV array are doing. The output of each is independent of the other modules/inverters in the set. The outputs of the microinverters or ac PV modules are connected in parallel, rather than in series, and this isolates one from another.

The outputs are at 120, 208, and 240 volts ac and these ac output circuits act much like ac branch circuits. They go dead when the ac utility power is removed at any disconnect in the circuit, so they do not pose the safety hazards associated with the daytime "always-energized" dc circuits operating at hundreds of volts between the modules and the inverter. If a short circuit or ground fault were to occur in these ac output circuits, the dedicated branch-circuit breaker would open and the circuit would go dead. Opening the main service disconnect or the backfed PV breaker will de-energize those PV ac output circuits and is one method of achieving compliance with 690.12 rapid shutdown requirements—a boon to fire fighters.

Disadvantages

There may be some cost impact of using ac PV modules or microinverters on each module when compared with using single-string inverters. However, two factors must be considered. The cost of dc switchgear and the required conduit (or other appropriate wiring method) for dc conductors inside a building, plus the cost of the single inverter, must be compared with the added cost of multiple small inverters or ac PV modules with an inverter on each module.

Then there are the life cycle costs. Modules are guaranteed for power production for 25 years, but can be expected to produce power for as long as 50 years. Large inverters generally have an average longevity of about 15 years. Microinverter manufacturers, using different construction methods and topologies, are predicting significantly longer lives for their products. However, the failure of a microinverter in a system or on an ac PV module may necessitate disassembling some portion of the PV array to gain access to the failed device. Time will reveal all.

DC-to-DC Converters

DC-to-dc converters are on the market. They are small boxes with leads that attach to the output of each module. These boxes act somewhat like a microinverter by decoupling the actions of one module from others in the string so that shading is less of a problem. Of course, the output of these devices is dc, not ac like the microinverter. Some of these devices can be used on just a few modules in a PV array, but others are required for every module. Some devices work with a standard string inverter, but others require a special inverter made by the same manufacturer. Most are connected in series like a normal string used with a string inverter.

In some cases, the specialized inverter required with some of these devices communicates and controls the performance of the individual dc-to-dc converters connected to each module. Removing the ac power from the inverter or turning it off can, in some cases, shut down the individual converters, removing dc voltages from the wiring.

These devices have a wide variation in system design and connection requirements. There are several sections of the *NEC* that address general requirements for dc-to-dc converters and they are found in Figure 690.1(a), 690.2, 690.4, 690.7(B), 690.8(A)(5) and (6), 690.9(A), 690.11, 690.15, and 690.53(3). The requirements are self-explanatory and are not associated with any particular product design or topology. However, because all of these dc-to-dc converters should be listed per 690.4(D), installers and inspectors are generally going to have to rely on instruction manuals and labeling supplied with these new devices for proper system design and installation [110.3(B)]. The concepts of short-circuit current and open-circuit voltages used in the design and installation of standard string module systems will not apply to these products.

Summary

Numerous microinverters and ac PV modules are being installed, and while they are directed at the smaller residential systems, systems as large as several megawatts have been installed using microinverters. They are being sold in home improvement centers, building supply houses, and local electrical supply houses and the public is buying them. Inspectors should become familiar with these devices and the *Code* requirements that apply to them.

53
54
55

05

Energy Storage Systems; Batteries in PV Systems

Article 706, Energy Storage Systems

Article 706, Energy Storage Systems (ESSs), was added to the 2017 *NEC*, and all references to batteries or other energy storage systems previously found in Article 690 are now in this section. This new section is quite lengthy and has requirements that were not previously in the *Code*, such as a requirement to determine and mark the rated short-circuit current of the ESS. Article 480, Storage Batteries, remains in the *Code* and has some of the same new or revised requirements that appear in Article 706.

Note that a PV system, as defined in Article 690, is separate and distinct from any energy storage system that it may be connected to. This separation, and the new grounding definitions and requirements for a PV system, may complicate the grounding requirements for both systems where they are combined. It will take time, and possibly the 2020 *NEC* will clarify this.

Section 690.71 states that an energy storage system connected to a PV system must be installed in accordance with Article 706, which applies to ESSs operating over 50 volts ac or 60 volts dc. It would appear that now only Article 480 applies to battery systems operating at below 60 volts dc, and this would include PV systems using 12-volt and 24-volt nominal battery systems and a few 48-volt nominal systems. However, some of the requirements, in Article 480, like for battery disconnects, apply only to systems operating over 60 volts dc. Installers of these low-voltage, but high-current, battery systems will have to exercise common sense in designing these systems, as will AHJs inspecting them. Article 720, Circuits Operating at Less Than 50 Volts (direct current or alternating current), is not going to help because it excludes PV systems in Article 690, Parts I and VIII.

> "**706.1 Scope.** This article applies to all permanently installed energy storage systems (ESS) operating at over 50 volts ac or 60 volts dc that may be stand-alone or interactive with other electric power production sources."

Energy Storage Systems Defined in Detail in the *Code*

"**Energy Storage System (ESS).** One or more components assembled together capable of storing energy for use at a future time. ESS(s) can include but is not limited to batteries, capacitors, and kinetic energy devices (e.g., flywheels and compressed air). These systems can have ac or dc output for utilization and can include inverters and converters to change stored energy into electrical energy.

> **"Energy Storage System, Self-Contained.** Energy storage systems where the components such as cells, batteries, or modules and any necessary controls, ventilation, illumination, fire suppression, or alarm systems are assembled, installed, and packaged into a singular energy storage container or unit.
>
> Informational Note: Self-contained systems will generally be manufactured by a single entity, tested and listed to safety standards relevant to the system, and readily connected on site to the electrical system and in the case of multiple systems to each other.
>
> **"Energy Storage System, Pre-Engineered of Matched Components.** Energy storage systems that are not self-contained systems but, instead, are pre-engineered and field-assembled using separate components supplied as a system by a singular entity that are matched and intended to be assembled as an energy storage system at the system installation site.
>
> Informational Note: Pre-engineered systems of matched components for field assembly as a system will generally be designed by a single entity and comprised of components that are tested and listed separately or as an assembly.
>
> **"Energy Storage System, Other.** Energy storage systems that are not self-contained or pre-engineered systems of matched components but, instead, are composed of individual components assembled as a system…
>
> Informational Note: Other systems will generally be comprised of different components combined on site to create an ESS. Those components would generally be tested and listed to safety standards relevant to the application."

This chapter will examine only chemical energy storage systems (batteries) that might be found in an off-grid, stand-alone PV system or in a residential or small commercial PV system with battery backup for part or all the loads. In many cases, these battery systems will be classified as an *Energy Storage System, Other*, because they will be field-assembled and installed at the final location.

Batteries in PV Systems

Electrical power outages are becoming more common in recent times with man-made and natural disasters and aging utility infrastructure. With natural disasters, such as Hurricane Sandy, tornadoes, and other severe weather conditions, many people who are already using PV systems (and many that do not have PV systems) are going to be interested in using PV systems in the event of electrical power outages. Electrical inspectors can expect to see increasing numbers of battery-backed-up, utility-interactive PV power systems.

PV Plus Batteries Means Power When the Utility Goes Out

These backup systems allow owners to operate some or all the loads in a building using a specially designed and configured PV system with batteries in the absence of the utility service (photo 5.1). These systems can be as small as a system that can power a radio or cell phone charger. They can also be as large as necessary to run all

Photo 5.1 • Battery-backed-up, utility-interactive PV system. DC coupled with PV array on the roof of a house.

appliances and loads in a residence or commercial building. The size and number of electrical loads that can be operated, and the period of time they can be operated, depends on the size of the PV power system, the size of the battery bank, and the size of the specialized inverter.

There are characteristics of PV systems with batteries that are different from those relating to standard utility-interactive PV systems. Obviously, batteries pose some unique problems that an inspector must review, and the connection of inverters to not only the electrical system in the house but also to the utility requires looking at some different *Code* sections than are normally used.

The multimode inverter that is used has characteristics of both the utility-interactive inverter and the standalone, off-grid inverter with features that are unique to the multimodal inverter. These inverters will be listed to UL Standard 1741. These inverters will have two sets of ac input/output terminals and a connection for the battery bank. Photo 5.1 shows the batteries and the multimode inverters in a system being installed.

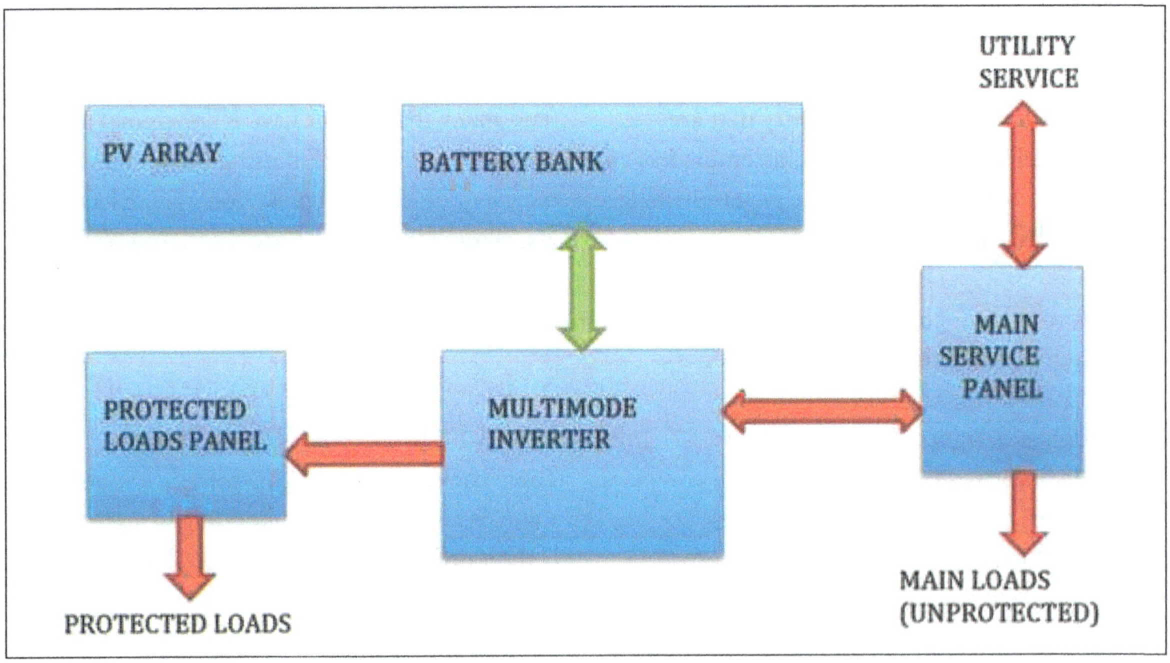

Figure 5.1 • Components in a battery-backed-up, utility-interactive PV system. The PV array may feed the batteries by two different means. See text and the following Figures.

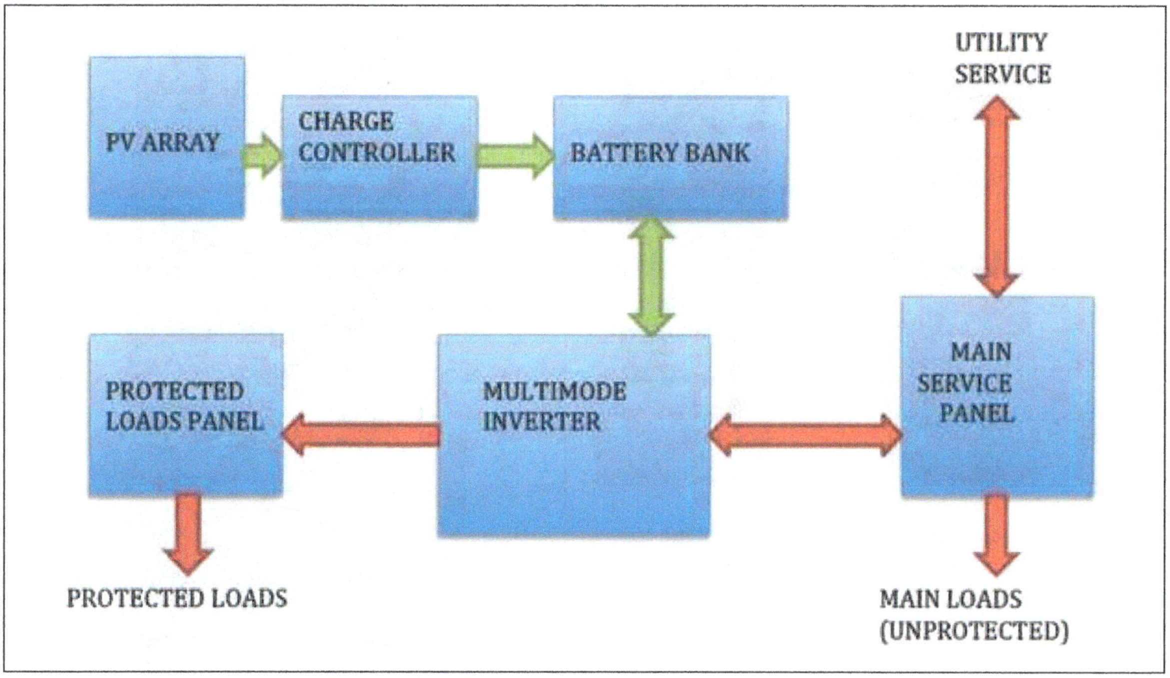

Figure 5.2 • DC-coupled system interconnections and power flows.

Figure 5.1 shows the basic elements of a battery-backed-up, utility-interactive PV system. Green arrows represent dc power/energy flow and red arrows represent ac power/energy flow. Double-headed arrows represent bidirectional power/energy flow.

DC-Coupled Battery Charging

There are two main types of battery-backed-up, utility-interactive PV systems. The first (and oldest) is what is called a dc-coupled charging system. As shown in Figure 5.2, the PV array has a nominal voltage of 24 volts or 48 volts and normally operates through a charge controller to charge a battery bank. The battery bank is connected to a multimode, utility-interactive inverter and that multimode inverter is connected to the house loads and to the utility using two separate and distinct ac input/output circuits. When the utility is present, the PV system charges the batteries through the charge controller. Power is taken from the batteries (or directly from the PV system when the batteries are fully charged) through the multimode inverter where it is converted to ac power.

The designated protected (backed up) loads may be supplied by either the utility (when present) or by the PV inverter output (supplied from the batteries when the utility is absent). Where the PV system power output exceeds the building loads, the excess energy is fed into the utility and renewable energy credits or net-metering benefits may be accrued. At night, or at other times when the PV production is low, power for the loads is purchased from the utility and fed to the main loads through the main panel or through the multimode inverter to the protected loads. In general, the battery stays fully charged at all times, but there are some systems in which the stored energy in the battery can be sent ("sold") to the utility with proper programming of the equipment.

When the utility is not present, the PV array and battery combination and the multimode inverter continue to operate the loads connected to the protected loads subpanel to the extent that the size of the PV system and the capacity of the battery bank can supply the energy required by those protected loads. The multi-

Photo 5.2 • DC-coupled system. Three black charge controllers on the right and four black inverters in the center. Silver enclosures contain ac and dc distribution equipment.

Photo 5.3 • 30-A charge controller with display. *Courtesy of Blue Sky Energy.*

mode inverter will not send power to the main (unprotected) loads or to the utility connection, but continues to monitor that utility connection for voltage and frequency. The main panel gets no power from any source. When the utility comes back online with the proper voltage and frequency characteristics, the multimode inverter will reconnect and the system becomes

Photo 5.4 • 60-A charge controller.

utility interactive once again. Photo 5.2 shows a dc-coupled battery charging system. The three charge controllers are on the right and the four inverters are in the center between the ac and dc distribution panels.

Charge Controllers

Standalone systems and dc-coupled utility-interactive systems with battery banks will also have charge controllers that regulate the state-of-charge of the battery bank. Charge controllers come in many sizes, shapes, and colors (photos 5.3, 5.4, and 5.5). When properly adjusted, they protect the batteries from being overcharged. Installers are responsible for adjusting these devices properly. Inspectors should verify good field terminations, proper conductor sizes, and appropriate overcurrent devices protecting those conductors.

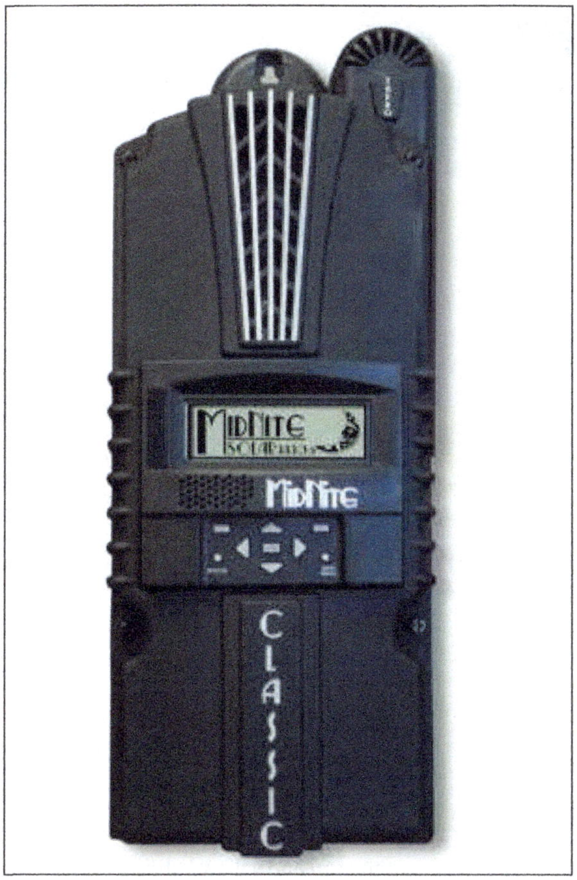

Photo 5.5 • 96-A charge controller. *Courtesy of MidNite Solar*

AC-Coupled Battery Charging

Figure 5.3 shows a more recent type of system, known as an ac-coupled charging system, where the PV modules are usually configured in a high-voltage string configuration (200 volts to 600 volts) and provide dc voltage to a standard utility-interactive inverter. The output of the utility-interactive inverter is connected to the protected load subpanel with a backfed breaker [705.12(B)] and that subpanel is connected to the load ac input/output terminals of the multimode inverter. The battery again is connected to the multimode inverter dc input/output. The utility is connected to its unique ac input/output on the multimode inverter, and when the utility is present, it feeds through the multimode inverter, generally keeping the batteries charged at all times and providing energy to the protected load subpanel. The utility-interactive inverter sees the proper voltage and frequency supplied by the utility and continues to convert dc PV energy into ac energy that can be used by the loads (both protected and main) and be fed to the utility.

When the utility goes down or has a brownout

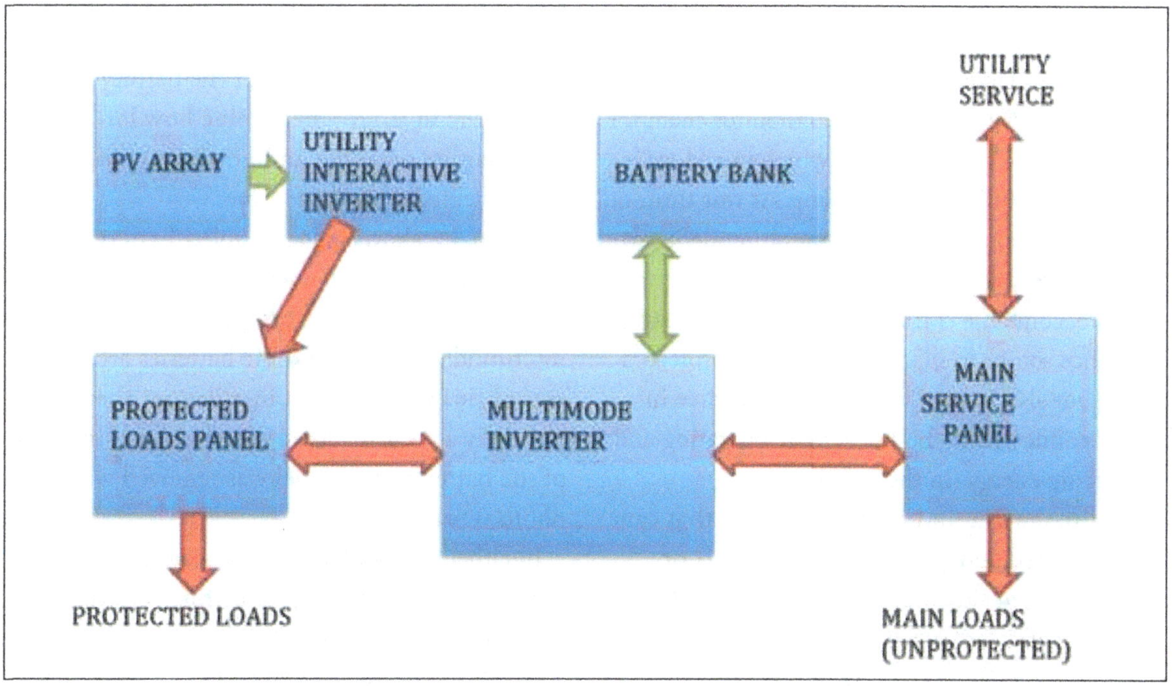

Figure 5.3 • AC-coupled system interconnections and power flows.

Photo 5.6 • AC-coupled system. The gray utility-interactive string inverters are at the top and yellow hybrid inverters below. The storage batteries are in green cases. The rotective plastic shield between batteries and electronic equipment is not shown.

(voltage or frequency variation), the multimode inverter senses this and stops sending power to the now unenergized utility lines (and the main load panel). However, it continues to monitor them for proper voltage and frequency, which would indicate that the utility is back online. At this time, on the load ac input/output of the multimode inverter, the battery supplies energy to the inverter and it will become the correct frequency and voltage reference source to supply not only the protected loads, but also to keep the utility-interactive inverter connected to the PV system, operating and producing energy (in the daytime).

Again, the size and number of loads that can be connected and operated for any short or long period of time depends on the size of the PV array and the capacity of the battery bank. Typically, the PV array may only supply energy for four to six hours per day. Loads obviously can operate 24 hours a day, so the total amount of PV array energy that can be stored in the battery—and the capacity of the battery and size of the inverter—determine how long the loads can be operated and how many loads can be connected at any one time.

Photo 5.6 shows an ac-coupled, battery-backed-up, utility-interactive system. The gray utility-interactive inverters are above the yellow multimode inverters and the batteries are in the rear of this very compact installation. There is normally a clear insulating service panel in front of the batteries; the panel was removed when the photo was taken.

In either case, with dc charging or ac-coupled charging of the batteries, the certified or listed multimode inverter ensures safety for the power line and utility personnel anytime the utility is shutdown or operates abnormally.

Battery Considerations

Batteries, although not considered a source of energy, can store considerable amounts of energy. They should not be considered current-limited sources as are PV modules, but, rather, they have the characteristics of a constant-voltage output, like an ac feeder with large amounts of available short-circuit current. Batteries must have overcurrent protection and disconnects on the output conductors. Batteries operating over 100 volts (conductor to conductor or conductor to ground) shall also have a maintenance disconnecting means for both ungrounded and grounded conductors that is accessible only to qualified persons. Where battery circuits are subject to field servicing and operating at over 240 volts (conductor to conductor or conductor to ground), they shall have provisions to disconnect the battery series-connected strings into segments that are less than 240 volts (706.21, 706.30). The current between the battery and the multimode inverter is bidirectional. It flows to the batteries when the batteries are being charged by the multimode inverter or the charge controller, and it flows from the batteries when the multimode inverter is in the inverting mode supplying the protected loads with ac power.

In the dc-coupled charging system, the conductors between the charge controller and the battery are sized based on the rated output of the charge controller, irrespective of the size of the PV system feeding it. These conductors should be sized at 125 percent of the rated output current of the charge controller [706.23(C)]. There should be an overcurrent device and a disconnect at the battery end of the circuit to protect these conductors from high short-circuit currents originating at the battery. Depending on the location of the charge controller with respect to other components, there may be disconnects required on the input and output of the charge controller. A main PV system disconnect located between the PV array and the charge controller will be required, complying with 690.13.

Available short-circuit currents. The battery banks used in these types of systems will typically have an available short-circuit current at the output conductors from the battery bank less than 15,000 amps. Conductor lengths, connections, and conductor resistances limit the available short-circuit current. Any overcurrent devices or disconnects must have ratings that can handle currents of this magnitude. Current-limiting fuses and dc rated circuit breakers are generally available with sufficient ratings and should be used.

Conductors. The conductors between the battery bank and the multimode inverter must carry bidirectional currents. The multimode inverter will use utility power or power from the utility-interactive inverter in ac coupled systems to keep the battery charged and currents will flow from the inverter to the battery. When the multimode inverter is operating in the inverting mode and supplying protected loads with energy, the currents will flow from the battery to the multimode inverter.

In general, the discharging currents flowing from the battery to the inverter will be larger than the charging currents flowing from the inverter to the battery.

Photo 5.7 • Class B stranded conductor, top; fine stranded conductors, bottom.

This is because the typical multimode inverter will be able to draw more current from the battery than it can provide to charge the battery. Therefore, the conductors between the batteries and the inverter must be sized based on the maximum rated output of the multimode inverter in the inverting mode of operation. However, if the multimode or stand-alone inverter instruction manual indicates a higher current in another mode of operation, then that current should be used. This continuous current (called the maximum current) should be specified in the inverter specification or installation manual and the conductor sized at 125 percent of this maximum current (706.20). Of course, the battery conductors should be in a raceway along with an equipment-grounding conductor, which would be used to ground any metallic battery rack and battery disconnect or overcurrent device enclosure. The size of the equipment-grounding conductor would be based on the rating of the overcurrent device protecting the circuit (250.122).

Many premanufactured battery cables are made with fine-stranded cables consisting of type appliance wire material (AWM) conductors. These cables are not suitable for use in battery PV systems because they are not mentioned directly in the *NEC* as one of the Chapter 3 wiring materials suitable for field-installed wiring. In many cases, automotive battery cables and welding cables have been used for their flexibility, but they are not allowed in this application by the *Code*. To address flexibility issues, see the fine-stranded cable requirements in Section 110.14.

THWN, RHW, THW and other commonly available conductors listed in Chapter 3 of the *Code* are acceptable for battery installations. Most commonly available plastic (thermoplastic) or rubber-insulated (thermoset) conductors are considered acid- and moisture-resistant.

However, the use of fine stranded flexible cables may pose termination issues. See *NEC* 110.14 and photo 5.7.

Photo 5.8 • Battery disconnect and overcurrent protection located near the batteries.

Battery Circuit Overcurrent Protection and Disconnects. An overcurrent device should be located at the battery end of the circuit to protect this conductor from high available fault currents from the battery. This overcurrent device will be sized at 125 percent of the standalone or multimode inverter rated dc current in the inverting mode, which is the same number used to size the conductors. An overcurrent device at the inverter end of the circuit is normally not required because the inverter typically cannot source the same high fault currents that the battery can. A battery disconnect should be installed at the battery end of the circuit. Normally, if the inverter is within four to five feet of the battery bank, it may not be practical or possible to put a disconnect any nearer to the battery than this distance. Therefore, the disconnect for this circuit can be near or at the inverter—usually in a power center. However, if the distance between the battery and the multimode inverter is more than four to five feet, or if the inverter is in a different room than the battery bank, then there must be a disconnect at the battery end of the circuit in addition to the overcurrent protection required at that location. Photo 5.8 shows a battery disconnect/overcurrent protection enclosure using circuit breakers mounted just above a valve-regulated (sealed) lead-acid (VRLA) battery bank. These batteries release no hydrogen gas or acid fumes during normal operation.

Grounding. The nominal battery voltage in these systems is 48 volts dc. The operating voltage may be as high as 62 volts to 65 volts. Normally, multimode inverters do not ground one of the battery circuit conductors and the grounding system used by the multi-mode or standalone inverter will have to be coordinated with the grounding system used by the connected PV system (690.41). Equipment grounding is required throughout the combined system and the requirements for this grounding have remained relatively unchanged in recent editions of the *NEC.*

If the system uses dc-coupled battery charging, the connection to earth for the PV system will usually be done through a distinct and separate ground-fault detecton interruption system (GFDI) as required by *NEC* Section 690.41. In some cases, the charge controller may have a GFDI built in.

On an ac-coupled system, the utility-interactive inverters will have their normal GFDI internal circuitry and will also be required to meet the requirements for PV system grounding in 690.41. However, in ac-coupled systems, the dc battery circuit may still have to be solidly grounded to keep costs down and to be compatible with available equipment that has been designed for use in grounded systems. Where an ungrounded battery system is used, there is no provision for having an overcurrent device in just one of the ungrounded conductors as there is for PV dc circuits. Overcurrent protection and disconnects must be used in both ungrounded conductors; therefore, increasing the cost of an ungrounded battery system.

AC Circuit Considerations

Multi-wire branch circuits. Many houses today have several multi-wire branch circuits that have two branch circuits with a shared neutral conductor and are wired with a 14–3 AWG/with ground type NM cable. Multimode inverters come with either 120-volt ac outputs or 120/240-volt ac outputs. Neither of these multimode inverters should be connected to load circuits in a building that are part of a multi-wire branch circuit. [See *NEC* 690.10(C).] The inverters in the inverting mode (standalone mode), in some cases, may not be in synchronization with the utility power frequency waveform. This could cause overloading of the shared neutral that is associated with multi-wire branch circuits. If any of the circuits needing battery backup power protection are multi-wire branch circuits, they should be segregated in their entirety (both circuits) in the special protected loads load center that is connected to the multimode inverter ac output.

Utility connections. One of the characteristics of most multimode inverters is that they can pass power from the utility through to the protected load circuits at a greater power level then they can supply power to the utility in the utility-interactive mode. This indicates that the circuit and overcurrent device (typically a breaker) between the utility connection and the multimode inverter must be rated at the full pass-through current capability of the inverter. A common value of this circuit breaker would be 60 amps or 70 amps. However, in the utility-interactive mode, the inverter may only be able to source 33 amps from the PV system into the utility. In previous editions of the *Code*, the 60-amp or 70-amp breaker would be used in the 705.12(D) calculations to determine panelboard/load center busbar ratings and conductor sizes. The danger to the circuit from overload-

Photo 5.9 • Flooded lead-acid batteries with acid containment system.

ing, however, is related to the 33-amp output of the inverter when feeding the utility.

Section 705.12(B)(2) now requires the calculations for this requirement to be based on 125 percent of the rated utility-interactive inverter output in the utility-interactive mode. In this example, 41.25 amps (1.25 x 33) could be used in the calculations. The circuit breaker connecting the inverter to the load center can still be rated at the higher 60 amps or 70 amps required to allow the protected loads to be operated in the pass-through mode of operation. Of course, the circuit conductors must be sized to carry the higher currents in the pass-through mode, including any ac currents from the utility used to charge the battery.

Aside from the battery circuits and the unique characteristics of the utility interconnection covered above, the multimode inverter in the battery-backed-up, utility-interactive PV system is connected to the utility in much the same manner as any normal utility-interactive system. The dc PV circuits are connected in the same manner as those circuits in a standard utility-interactive PV system for the ac coupled system. The dc-coupled systems require additional considerations for the low-voltage battery charging circuits.

A Closer Look at Batteries

Energy storage systems (in the form of batteries), when included in photovoltaic power systems, are critical and important items that need close scrutiny during the plan review and inspection process.

Types of Batteries

There are two main types of battery systems that have been used (and are still being used) in PV systems at this time. Both are based on a lead-acid battery technology (older than the *NEC*).

The oldest technology uses a flooded lead-acid battery that has removable vent caps and requires the addition of water on a regular basis (photo 5.9). A related technology known as the valve-regulated, lead-acid (VRLA) battery uses an internal chemical design to allow the battery case to be effectively sealed (photo 5.10). No water can be added to VRLA batteries.

The newer battery types of nickel cadmium and lithium ion have not found their way into PV systems in any substantial numbers yet, probably due to the higher cost of these technologies. Increasing use of these battery types in electric vehicles is lowering the costs and they are being used more frequently in PV applications. (See the AC Battery Section below.) Most lithium ion battery systems require a charge management system for each battery cell and that system increases the cost substantially. Other advanced technology battery types, such as chemical "flow" batteries, are generally found only in large, experimental installations in utility and industrial applications.

Mechanical Installation, Venting, and Acid Containment

Batteries contain substantial amounts of lead, which makes them very heavy. The floor that they rest on must be sufficiently strong, particularly for larger battery banks that may weigh several tons. Where battery banks are installed in racks that are more than a foot or so high, those racks must be secured to a substantial wall. Building codes, especially in earthquake zones, require substantial mechanical securing of batteries.

During the normal charging process, flooded lead-acid batteries will emit hydrogen gas and sulfuric acid fumes into the surrounding environment. The type and adjustment of the charge controller will determine the amount of outgassing. Although VRLA batteries are sealed, if the charge controller is misadjusted, becomes defective, or fails, VRLA batteries will also outgas

Photo 5.10 • Valve regulated lead-Acid (VRLA) batteries.

hydrogen and sulfuric acid fumes. Both battery types should be installed in a well-ventilated area and should normally not be installed in a living space.

In normal operation, the amount of hydrogen gas and sulfuric acid fumes are very limited and are quickly and easily dispersed into the surrounding room without problems. However, if these gases are restricted from being diluted with the surrounding air, an explosive combination of gases and air is possible. Garages and outbuildings are ideal locations for these batteries if they are well ventilated, not sealed, not used as living spaces, and do not have living spaces over them.

No attempt should be made to construct a manifold or venting system for these batteries. Power venting systems have a tendency to fail and the high-vent, low-vent system used for combustion appliances, such as gas furnaces and water heaters, is not applicable to batteries because there is no driving energy source equivalent to the heat energy source of a heating system.

Power-vented manifolds on the batteries should not be used, as they have been responsible for explosions and fires in the past. Batteries installed in well-vented locations are acceptable. Catalytic battery vent caps, employing a platinum catalyst, are available for use on flooded lead-acid batteries to reduce hydrogen outgassing and reduce the requirement to add water to the battery.

Flooded lead-acid batteries contain liquid

Photo 5.11 • Flooded lead-acid batteries in lockable polyethylene containers.

sulfuric acid, and VRLA batteries include a similar jelled electrolyte (photo 5.9). These batteries should be installed in a manner that prevents them from being mechanically abused. Many local building codes require some sort of acid containment for these batteries should the cases become cracked. Installing batteries in an acid-resistant outer container may be sufficient. Note the acid containment system under the batteries in photo 5.8. There are no generally available, certified or listed battery containers or battery boxes. For smaller systems, it has been found that heavy-duty polyethylene toolboxes provide sufficient acid containment and also (with lockable tops) meet *Code* requirements to prevent unqualified people from accessing the electrically energized battery terminals [110.27, 706.30]. (See photo 5.11.)

Location, Location, Location

In small- and medium-sized PV systems, batteries operate at a nominal 12, 24, or 48 volts. To minimize voltage drop in the conductors at typical high operating currents, batteries are generally installed as close as possible to the loads that they serve. In PV systems, these loads are usually a multimode utility-interactive inverter or a standalone inverter. In many cases, particularly in standalone systems, a power center is used that provides a central location for overcurrent protection, disconnects, and charge-control functions. Even in the best of circumstances, it is generally not possible to mount and connect an overcurrent device or a disconnect any closer to the battery terminals than three or four feet.

Conductors must be sized based on the continuous current requirements, which will be based on charging currents from either the PV array or an inverter/charger or on discharging currents, usually to the inverter. Depending on the system design, there may be more than one circuit connected to the battery—one circuit for charging, and one circuit for discharging. Where there is a single circuit used for both charging and discharging, the conductor must be sized on the largest continuous current in either mode. Although the *Code* typically requires that conductors be sized and protected based on continuous loads (3 hours or more), battery-based systems may require a slightly different approach.

With standalone, off-grid inverters, and multimode inverters in the standalone mode, inverters have a substantial capability to surge currents above their steady-state rating for periods of minutes to an hour. This may create voltage drop in the conductors that poses operational problems. In this case, voltage-drop calculations must also be made to ensure that the inverter will operate properly under all conditions of continuous use and surges.

Battery banks connected with sets of series-connected cells or batteries and parallel sets of cells or batteries should be wired so that the length of conductors (and the resulting resistance) would be the same for each series connection to minimize charging and discharging current and voltage imbalances and premature cell or battery failure. (See figure 5.4.)

Overcurrent Protection

Batteries can source high fault currents. It is somewhat difficult to obtain specific short-circuit current data on batteries and even more difficult

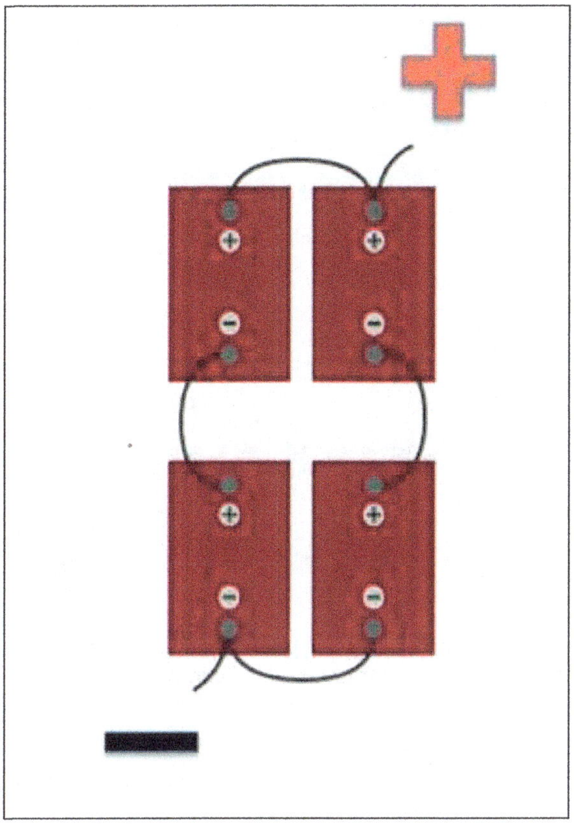

Figure 5.4 • Series-parallel battery connections. Note: Current travels through equal lengths of conductors in each of the series paths. Battery conductors should be the same size and length in each path. All terminations should be identical. *Source: Trojan Battery Co.*

to determine what those short-circuit currents may be at the output of the battery bank, which may consist of numerous series and parallel connected individual batteries.

Tests and estimates made several decades ago indicate that for typical residential and small commercial battery banks, the inter-cell/inter-battery connectors, the battery conductors, the contact resistances, and other factors will typically limit the available short-circuit current at the output of the battery bank to less than 15,000 amps. Therefore, any overcurrent device protecting these conductors should have an interrupt rating of at least 15,000 A. Both dc rated fuses and dc rated circuit breakers with that rating are available.

Direct-current-rated, current-limiting fuses are commonly available and will provide some protection for downstream components. Direct-current-rated circuit breakers, on the other hand, have little or no current-limiting capability. Any disconnecting equipment in the dc battery conductor path should have sufficient short-circuit interrupt capabilities where circuit breakers are used.

Normally, in small residential and small commercial battery banks, the individual strings of batteries (series-connected cells or batteries) are not fused. Series and parallel connections are made within the battery bank and then the overall output is provided with overcurrent protection. This would generally indicate that the conductors within the battery bank be able to handle the entire current of the battery bank (and not be sized to handle just a proportional part of the total battery current).

In parallel-connected sets of batteries or cells, the individual cells and batteries may not age at the same rate and the currents may tend to be higher in some of the series connections than in others. For that reason, it is somewhat difficult to fuse individual strings of batteries because they may be required to carry more current as other strings of batteries age and carry less current. Also, installing an overcurrent device inside the battery bank on a single string or several strings poses both maintenance and potentially explosion hazards.

In larger battery installations (where room-size battery banks are involved), individual strings of batteries or cells may have a fused disconnect on each string. These fused disconnects must be located in areas that allow easy access and are not subject to hydrogen gas or sulfuric acid fumes. The size of the conductors for each battery string and the overcurrent device rating require careful consideration because of the potential for unequal currents in the battery series circuits as the battery bank ages.

Proper maintenance of the battery banks is the

key to maintaining battery longevity and keeping the various currents balanced. However, aside from access requirements (706.33), the *NEC* does not address the maintenance of electrical systems.

The batteries are able to generate more short-circuit current than the typical charging source and there should be an overcurrent protection device near the batteries. The overcurrent device must be sized at least 125 percent of the continuous current (either charge or discharge, whichever is greater, including any long-term surge currents). If the charging currents are greater than the discharging currents, an overcurrent device may be required at the charging source.

A disconnect may be required at the battery to not only service the fuse (a circuit breaker includes the disconnect), but also to provide a method of disconnecting the battery circuit from other connected devices. Sections 706.30 and 706.31 establish disconnect requirements and cell interconnections.

Fuses that bolt directly to the battery terminals are generally not acceptable in *Code*-compliant installations because they are very difficult for unqualified people to service on the always-energized battery terminals. Also, the fuses may be in a hydrogen gas/acid fume environment under certain circumstances, resulting in OCPD failure or an explosion should the fuse open in that environment.

The AC Battery

The latest addition to energy storage systems is the ac battery (photo 5.12). No, this is not a contradiction in terms. This system was designed to provide a degree of energy backup during utility outages; to store PV energy for use when utility rates are high; and to prevent PV energy

Photo 5.12 • AC battery system. *Courtesy of Enphase*

from backfeeding the utility. These devices are quite complex (installation, set up and operation manuals may run hundreds of pages), but are usually a single piece of equipment with one or more external current sensors. The device includes a multimode inverter and batteries—usually some form of lithium-ion cells. The *NEC* has little information on the installation of these listed units and they must be installed and operated per the instructions provided with the listed unit [110.3(B)]. These AC batteries are made by at least three companies with more on the way.

Summary

Plan reviewers and inspectors should check battery installations for substantial mechanical installation; correct conductor types and sizes; proper placement and rating of overcurrent devices and disconnects; good conduit installation techniques; and proper ventilation.

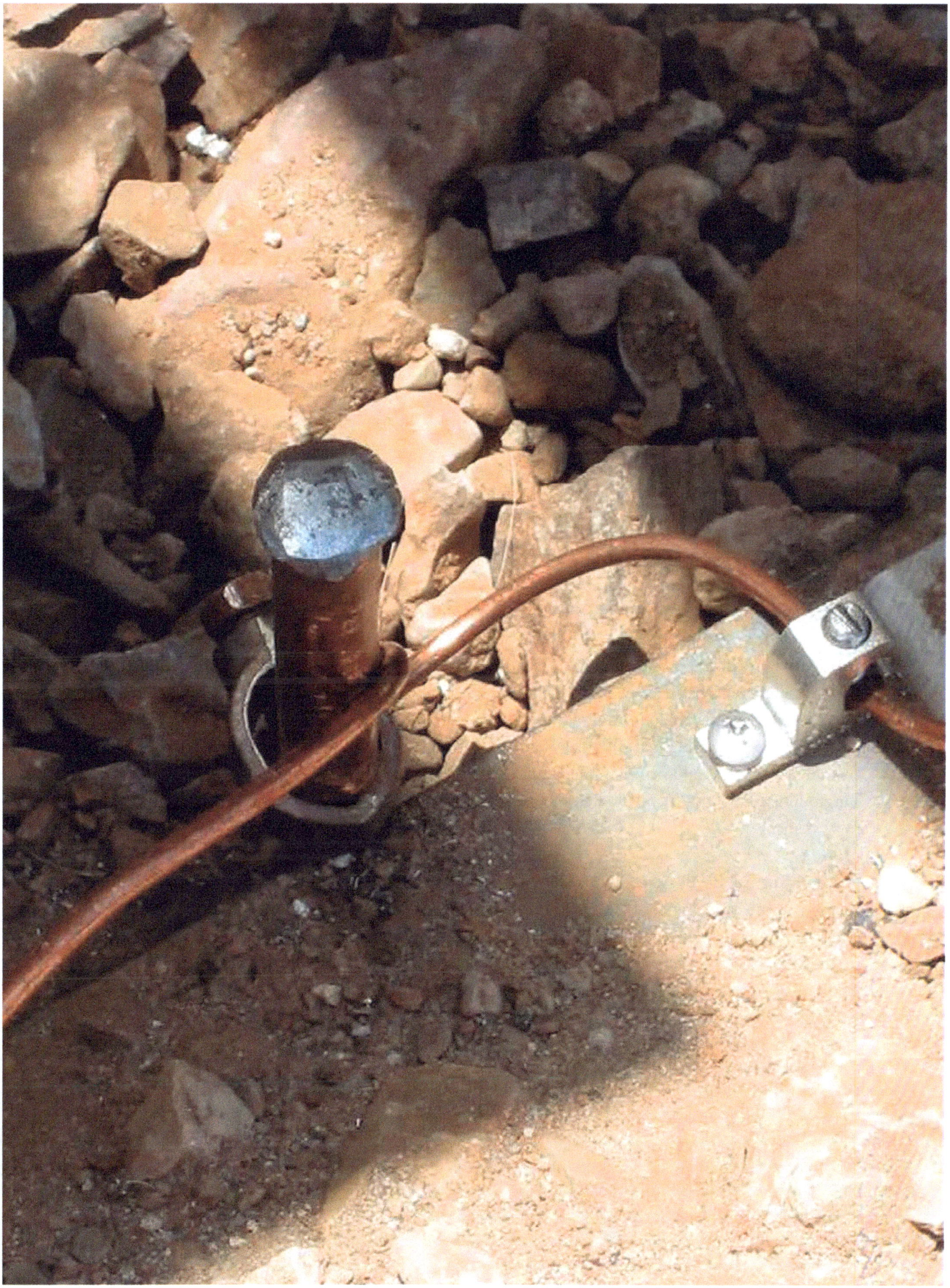

06

Grounding, Disconnects, and Overcurrent Protection

Grounding

Grounding—New and Old

Grounding PV systems has remained essentially unchanged for the most part since PV came to the *Code* in 1984. But with the 2017 *NEC*, the terminology that is used to describe various grounding functions has been changed. This was done to reduce confusion and simplify terminology between older "Grounded PV Array/Isolated Inverter" PV systems and newer, more common "Ungrounded PV Array/Non-isolated Inverter" PV systems.

Grounding the PV System Circuit Conductors. In the early years of PV (up to about 2005), we primarily had grounded PV arrays and grounded dc battery systems where one of the circuit conductors was "grounded" or "connected" to the grounding system. This system is composed of the equipment-grounding system, the grounding electrode conductor, and the grounding electrode system.

In some cases, the "grounding" method was a solid conductor, and in other cases, the grounding method was by a fuse or circuit breaker like those used in Section 690.5 of the 2014 *NEC*, Ground Fault Protection Device. In other cases, various resistances or solid-state devices were used to "ground" one of the circuit conductors.

In recent years, there are increasing numbers of "ungrounded" PV arrays and non-isolated (transformerless) inverters which have no dc circuit conductors connected directly to ground. Grounded PV systems in any but the very smallest PV systems are becoming a thing of the past, but will exist for years [690.41(A)(5)]. The *Code* requirements to address these varying "grounding" systems were cumbersome, difficult to understand, and hard to apply in many cases. The changes in the *2017 NEC* address and simplify many of these grounding requirements. However, it is noted that proposals for the 2020 *NEC, if adopted*, may further clarify these new requirements.

Some of the grounding definitions in the 2017 *NEC* are the same as in the *2014 NEC*.

From Article 100

"**Ground.** The earth.

"**Grounded (Grounding).** Connected (connecting) to ground or to a conductive body that extends the ground connection.

> **"Grounded, Solidly.** Connected to ground without inserting any resistor or impedance device.
>
> **"Grounded Conductor.** A system or circuit conductor that is intentionally grounded."

From Article 690

"Functional Grounded PV System. A PV system that has an electrical reference to ground that is not solidly grounded.

> Informational Note: A functional grounded PV system is often connected to ground through a fuse, circuit breaker, resistance device, non-isolated grounded ac circuit or electronic means that is part of a ground-fault protection system. Conductors in these systems that are normally at ground potential may have voltage to ground during fault conditions."

As an aside, my high school English teacher would have objected to the term *functional grounded* and preferred *functionally grounded*, but then she has not seen the rest of the *NEC*. *Functional ground* would probably have been acceptable—another potential update for the 2020 *NEC*.

> **"690.41. PV System Grounding Configurations**
> One of the following system grounding configurations shall be employed:
> (1) 2-wire PV arrays with one functional grounded conductor.
> (2) Bipolar PV arrays according to 690.7(C) with a functional ground reference (center tap).
> (3) PV arrays not isolated from the grounded inverter output circuit.
> (4) Ungrounded PV arrays.
> (5) Solidly grounded PV arrays as permitted in 690.41(B) Exception.
> (6) PV systems that use other methods that accomplish equivalent system protection in accordance with 250.4(A) with equipment listed and identified for the use."

It should be noted that there are no voltage requirements associated with the requirements in 690.41 and that old 50-volt limit is gone. It will take some time to figure out what we call these various types of PV arrays. For example, a PV array connected to a non-isolated inverter with a 240 VAC output will meet the requirements of (3) because at some point there is a utility transformer that has a grounded 240/120-volt center tap.

Equipment Grounding

Sections 690.43, 690.45, and 690.46 dealing with equipment-grounding conductors have remained essentially unchanged.

Grounding Electrode Systems

Section 690.47 on grounding electrode systems has been substantially revised and clarified. The most significant change is for functional grounded pv systems (not solidly grounded). In these systems, the normal ac equipment-grounding conductors from the inverter to associated distribution system are allowed to serve as the grounding connection for any ground-fault protection equipment and for the equipment grounding of the PV array. This is a change from previous 690.47(C)(3) in the *2014 NEC* that required this conductor to meet the requirements of what was called a "grounding conductor" (bonding, size, and continuity), but was actually a bonding jumper, as well as the requirements for an equipment-grounding conductor.

Grounding the Array—690.47(D) Grounding Electrode System

Section 690.47(D) has been revised in every edition of the *Code* since it was originally written.

Here is what is says, in part, in the 2017 *NEC*:

> "(A) Buildings or Structures Supporting a PV Array. A building or structure supporting a PV array shall have a grounding electrode system installed in accordance with Part III of Article 250."

This grounding electrode system could be the existing grounding electrode system or a newly installed one if no grounding system exists.

> "PV array equipment grounding conductors shall be connected to the grounding electrode system of the building or structure supporting the PV array in accordance with Part VII of Article 250. This connection shall be in addition to any other equipment grounding conductor requirements in 690.43(C). The PV array equipment grounding conductors shall be sized in accordance with 690.45."

Section 690.43(C) requires that these PV array and mounting structure equipment-grounding conductors be routed with the circuit conductors where those conductors leave the vicinity of the PV array.

690.47(B) permits, but does not require, additional auxiliary grounding electrodes for ground-mounted and roof-mounted arrays. They are to be installed in accordance with 250.52 and 250.54, which do not require them to be bonded to other grounding electrodes in the system. Of course, they would be connected together with the other grounding electrodes through the equipment grounding conductors.

Also, any building or structure that supports a PV array must have a grounding electrode system (photo 6.1)—either an existing one or an added one. The PV array equipment-grounding conductors will be connected to that grounding electrode system, and that grounding electrode system may be the existing site grounding electrode system if one exists. These PV array equipment-grounding conductors are routed with the dc circuit conductors.

Solidly grounded systems will require a grounding electrode conductor per Section 250.166 connected to the grounding electrode system. The grounding electrode conductor or main bonding jumper will be connected at any point on the PV output circuit (690.42).

Photo 6.1 • Array grounding electrode required. A 2 AWG copper conductor attached to a rebar to makes a concrete-encased electrode (sometimes referred to as UFER)—prior to concrete pour.

Disconnects

The various sections of the *Code* dealing with disconnects in PV systems have been significantly revised in the 2017 *NEC*. As noted in chapter 1; *NEC* Figure 690.1(B); and as shown in Figures 1.22, 1.23, and 1.24, the PV system disconnect can be either ac or dc and is the demarcation between the PV system and any other source of energy including the utility grid, energy storage systems, and other energy sources like fuel cells or wind systems. Section 690.13 deals primarily with the PV system disconnect, and Section 690.15 establishes requirements for the disconnection of PV equipment (photo 6.2).

Evidently through the use of master keys and fire axes, fire services consider the interior of a locked home to meet the readily accessible requirement, because about half of the homes in the United States have the main ac service disconnect located inside the structure. In the future, some consideration should be given to the differences

Photo 6.2 • Disco Madness. How many disconnects can you find? At least nine are visible in this 2014 *NEC* installation, but how many are inside the enclosures? And what would it look like under the 2017 *NEC*?

Photo 6.3 • There is no longer a specific requirement to have a dc PV disconnect located at or near the point of entry of the conductors to a building. Marking requirements have changed from this 2011 *NEC* commercial installation.

between an ac service and a dc PV circuit entering a building with respect to disconnect requirements.

No Longer at the Point of Entry. Section 690.13(A) requires that the PV system disconnect be in a readily accessible area, and this may be a dc disconnect in some systems or an ac disconnect in others [See *NEC* Figure 690.1(B)]. Note that there is no longer a specific requirement for a dc PV disconnect to be located inside or outside nearest the point of entry of the dc conductors into the building as required by the 2014 and previous editions of the *NEC*, such as 690.13(A) in the 2014 *NEC* (photo 6.3).

While there is a requirement in 690.15 in the 2017 *NEC* for an isolation device within 10 feet of the inverter, that device may not be located at the point of entry of the dc conductors if the inverter is more than 10 feet from that point of entry. Of course, the dc conductors inside the building are still required to be in a metal raceway or installed in Type MC cable [690.31(G)], but there is no longer an exception in 690.13 related to that installation requirement that allows nearly any location for the PV system disconnect.

An Informational Note to 690.13(A) alludes to the 690.12 rapid shutdown system as dealing with energized conductors entering the building, but the author feels that this is a separate system/function that is not directly related to the widespread requirement of having all power sources equipped with a disconnect as they enter a building [230.70(A)(1)]. This issue may or may not be clarified by the 2020 *NEC*.

While some AHJs and jurisdictions may continue to follow the 2014 *NEC* guidance in this area, the 2017 *NEC* apparently allows or accepts the required dc isolation device/disconnect on the dc input of the inverter or the input to a charge controller as a suitable dc PV disconnect, no matter where it is located. As long as that location is readily accessible, the disconnect can also serve as the PV system disconnect where it meets the definition and requirements of that device. In some cases, the disconnect may be the PV system disconnect as defined in the 2017 *NEC*. In other cases, it may just be a dc disconnect at the input to a utility-interactive inverter.

Various parts of 690.13 require that the PV

Photo 6.4 • Disconnect labels based on 2014 *NEC*. 2017 *NEC* requirements are slightly different.

system disconnect open all conductors [690.13(F)]; be rated for the available short-circuit current and voltage in the circuit [690.13(E)]; be rated as service equipment for ac supply-side PV connections [705.12(A), 690.13(C)]; and may be a single disconnect for multiple inverter systems and have up to six switches for more complex PV systems [690.13(D)]. As mentioned previously, the color coding of a solidly grounded conductor that has a disconnect will need to be clarified in the *Code*.

To Open the Grounded Conductor or Not? Because the PV system disconnect may be the connection of the output of a utility-interactive inverter to the utility power source, the requirements of disconnects and Article 690 should be consistent with those requirements in Article 705. The requirements in 690.13(F) require that the PV system disconnect simultaneously open *all* conductors of the PV circuit connected to other sources. There is no exception for grounded conductors and it is noted that, in many cases where a grounded conductor is involved (usually a neutral) in that circuit, at least a three-pole disconnecting means would be required. It will be difficult to achieve the ac PV system disconnect using the backfed circuit breaker in a load center due to that neutral switching requirement. On the other hand, Section 705.21 requires that only the ungrounded conductors be disconnected at this point.

Proposals have been made to correct the requirement in 690.13 in the 2020 *NEC*, but in the meantime, it is suggested that the requirements of 705.21 be followed rather than those of 690.13(F). A Tentative Interim Amendment (TIA) for the 2017 *NEC* may rectify this issue before the 2020 *NEC* is published and adopted.

Isolation Devices—a Subset of Disconnecting Means

Section 690.15, Disconnection of Photovoltaic Equipment, requires that isolating devices for all non-solidly grounded conductors be located within 3 m (10 ft) of PV modules, ac PV modules, fuses, dc-to-dc converters, inverters, and charge controllers.

Note the change from previous editions of the *Code*. The 2014 *NEC* in 690.13(F) specifically excluded the requirement for a PV dc disconnect at either the module or the array. That exclusion does not appear in the 2017 *NEC*, so an isolation device (circuits of 30 amps or less) or a disconnecting device (circuits more than 30 amps) will be required at the PV modules (and ac PV modules, etc.) and possibly the PV array. While the PV module (ac PV modules, dc-to-dc converter) connectors could be used as module/converter isolating devices, in many installations these connectors are neither visible or accessible after the PV array is installed on the roof. But visibility or accessibility is not a *Code* requirement for these isolation devices. Although a disconnecting means on the output of a combiner (where used) is no longer specifically required [see 2014 *NEC*, 690.15(C)], such a disconnect/isolation device as required by 690.15 might meet the PV module isolation device requirement at the PV array.

A disconnecting means can always be used as an isolation device, but for circuits with a maximum of 30 amps or less, the isolating device may be a connector (690.33) if listed and identified with a specific piece of equipment; a finger-safe fuse holder; an isolation switch that requires a tool to open; or an isolating device listed for the intended operation. These types of isolating devices shall not be required to have an interrupt rating, nor shall they be required to open all conductors of a circuit simultaneously. They will be able to open the maximum circuit current or be marked "Do Not Disconnect Under Load" or "Not for Current Interrupting."

It is not clear how the disconnecting means marking requirements of 690.53 will be met when those disconnecting means are connectors.

For circuits with a maximum current greater than 30 amps, an equipment disconnecting means must be used to isolate the components. The equipment disconnecting means must open all conductors of a circuit simultaneously and have an interrupt rating sufficient for the available short circuit current and voltage at its terminals [690.15(B)]. Section 690.15(D) lists the types of devices that are suitable for use as equipment-disconnecting means. These devices include:

"(1) A manually operable switch or circuit breaker

(2) A connector meeting the requirements of 690.33(E)(1)

(3) A load break fused pull out switch

(4) A remote-controlled circuit breaker that is operable locally and opens automatically when control power is interrupted."

Equipment disconnecting means that may have both line and load terminals energized when open shall be marked with the electric shock warning statement in 690.13(B).

Photo 6.4 shows the front of a PV dc disconnect with the labels required by 690.13(B) and 690.53 (2014 *NEC*). The 690.13(B) warning is required because the load terminals of this disconnect are connected to the inverter dc input, which may be energized for up to five minutes after the disconnect has been opened. The filter and energy storage capacitors in the inverter will be discharged after this time. The 690.53 label with the system dc voltages and currents will allow the AHJ to determine if the correct conductors, disconnects, and overcurrent devices have been installed. Section 690.53 in the 2017 *NEC* requires a revised set of information:

(1) Maximum voltage
Informational Note to (1): See 690.7 for voltage.

(2) Maximum circuit current
Informational Note to (2): See 690.8(A) for

Photo 6.5 • Line terminals at top and protected with shield; unprotected load terminals at bottom.

calculation of maximum circuit current.

(3) Maximum rated output current of the charge controller or dc-to-dc converter (if installed).

Power flows in a PV system from the PV array through the dc PV disconnect, the inverter, the ac disconnect, various isolation devices, and finally to the grid. This power flow sometimes confuses installers on how to properly connect the dc and ac disconnects. Note the upper line-side terminals on the disconnect shown in photo 6.5 are covered by an insulated cover. Also note the switchblades, the fuse holder terminals (if any), and the load-side lower terminals are exposed and easily touched.

A general safety rule is that the most dangerous circuit should be connected to the protected, line-side terminals. If this is done, it is less likely that energized terminal connections will be accidentally touched when the door of the disconnect is open. In the PV dc disconnect/isolator, the PV source or output circuits should always be connected to the line-side terminals where a switched disconnect is used.

The dc input to the inverter is connected to the load-side terminals and the 690.13(B) warning label is required as shown in photo 6.4. The dc input terminals are connected to energy storage capacitors inside the inverter and may be energized for up to five minutes. On the ac disconnect, the utility connection goes to the protected line-side terminals at the top.

Marked "Line" and "Load." To Backfeed or Not? AHJs have noted that many of the utility-required ac PV disconnect switches (visible blade, lockable open) have terminals marked "Line" and "Load," but are being backfed because the utility connection should be made to the touch-shielded, protected upper "Line" terminals (photo 6.5). Section 690.13(F)(2) states:

> "**(2) Devices marked "line" and "load."** Devices marked with "line" and "load" shall not be permitted for backfeed or reverse current."

This section appears to confirm the requirement that switches (as devices) so marked should not be backfed (photo 6.6). However, Section 705.12(B)(4) provides slightly different guidance:

> "**(4) Suitable for Backfeed.** Circuit breakers, if backfed, shall be suitable for such operation.
>
> Informational Note: Fused disconnects, unless otherwise marked, are suitable for backfeeding."

Discussions with personnel at UL add additional weight to the Informational Note and confirm that there are no issues with backfeeding fused or unfused switched disconnects unless they are specifically marked "Not suitable for back feeding" or the

Photo 6.6 • Note the factory embossed "LINE" marking on this unfused ac disconnect (safety switch) just below the word "UTILITY" on the field-installed label. Okay for backfeeding with the utility connection to the line terminals.

equivalent. The "Line" or "Line and Load" markings on these switches are used to identify the shielded, protected terminals where the utility (or other most likely energized circuit when the switch is opened) is to be connected. Proposals for the 2020 *NEC* should clarify this area.

Photo 6.7. AC and dc connectors on this microinverter are certified/listed as load-break-rated disconnects.

Utility-Interactive Inverters on the Roof

In installations using microinverters on the roof, or installations where the string inverters are located on the roof of the building, *NEC* 705.70 has specific disconnecting requirements related to those types of installations where the inverters are in not in readily accessible areas. While this section is self-explanatory, some additional information is needed when it comes to microinverters and ac PV modules. To meet the disconnect/isolation requirements at the inverter location, most microinverter companies have had their ac and dc connectors listed as load-break-rated disconnects when used in this application. This information should appear in the instruction manual for the particular product and is covered in the *Code* by 690.15(C)(1).

Photo 6.8a • An 8-kW string inverter can be paralleled with two others on a special combining cable and using a custom splicing device with no OCPD for each inverter. *Courtesy of HiQ Solar.*

(See photo 6.7.)

AC PV Modules and Microinverters

In recent times, ac PV modules and microinverters are the only types of inverters that are paralleled at their outputs on a common circuit (called the trunk cable) without a dedicated overcurrent protection device for each inverter. As the power rating of these devices increases (currently up to about 700 watts), some more precise definition of these inverters will be needed in the *Code* because the power levels of these smaller microinverters are approaching the "string" inverter power levels. There are now three-phase, 480-volt ac string inverters rated at 8 kW that have been listed for parallel operation of three inverters with no overcurrent protection for each inverter when used with a listed combining "cable" and where specially listed ac connectors are used as the output disconnect for each inverter (photos 6.8a and 6.8b).

Because the ac connectors are listed as load-break connectors at each microinverter and these newer string inverters, there is no requirement for an additional ac disconnect for the trunk cable on the roof. In many cases, several trunk cables (or their extensions using an appropriate wiring method) from sets of microinverters or ac PV modules are run from the roof to an ac inverter combining subpanel in a readily accessible area where each of the backfed breakers for each of the sets of microinverter outputs serves as the second ac disconnect required by 705.70.

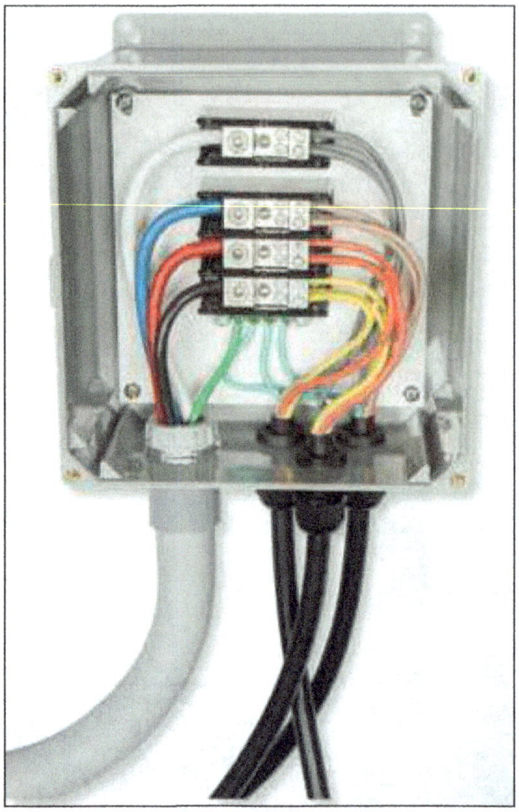

Photo 6.8 b • Custom splicing device used to parallel three 8-kW utility-interactive inverters with no OCPD for each inverter output. *Courtesy of HiQ Solar.*

Standalone PV Systems

In the larger standalone systems that are used to power dwellings or other structures, the ac output of the inverter is normally fed to a power distribution panel. This power distribution panel should have a main breaker, and that breaker is equivalent to the main ac service disconnect on utility-connected electrical systems. *NEC* Article 710 provides limited guidance on the requirements for disconnects and their location. If the inverter is in the interior of the house, there appears to be no specific location for the main

Photo 6.9 • (*left*) PV rapid shutdown initiator located near incoming service entrance conductors and readily identifiable by the fire service.

Photo 6.10 • (*top right*) Inverters with attached dc disconnects. AC disconnects are in separate nearby panelboards.

ac disconnect. However, AHJs who inspect such systems sometimes consider requiring an external ac disconnect on the dwelling or at least a directory on the outside of the building indicating the location of the ac disconnect inside the building. When the inverter is located in a separate structure or is exterior to the dwelling, then the normal *Code* rules for installing an ac disconnect for the conductors entering the building or structure would apply [230.70(A)]. In some jurisdictions, the fire service may have established requirements for disconnects on standalone PV systems.

The currently evolving changes in the *Code* and in UL standards. *NEC* 690.12, Rapid Shutdown System for PV Systems, will have some effect on how the PV system can be controlled from a single point in an emergency situation (photo 6.9).

Many utility-interactive inverters have an internal dc disconnect or one in an enclosure that is attached to the inverter. That disconnect will serve, in most cases, as the dc isolator for the inverter (690.15) (photo 6.10).

Disconnect requirements with respect to location vary significantly with each particular PV installation. Those requirements are being modified for clarity and are addressing new equipment in each new edition of the *NEC*. To paraphrase the Bard again: We know what disconnects are, but know not what or where they may be.

PV Circuit Overcurrent Protection

Section 690.9 establishes the requirements for overcurrent protection associated with the now redefined PV system circuits, both dc and ac. Overcurrent protection requirements for batteries (energy storage systems), stand-alone PV systems, and dc and ac microgrids are covered in other articles in the *Code* and in other chapters in this book.

Sometimes overcurrent protection is not needed. In any situation where circuit conductors have sufficient ampacity for the sum of all available currents in that circuit under normal or fault conditions, overcurrent protection is not required [690.9(A)].

Special Consideration and Location. Although circuits are normally protected from overcurrents at their source, PV source and output circuits and interactive inverters differ from that general rule. Because several pieces of PV equipment, such as PV modules, dc-to-dc

Photo 6.11 • Fuse listed for PV applications.

converters, and interactive inverters, have current-limited outputs, circuits connected to those devices require special consideration with respect to overcurrent protection. These circuits, when connected to a source that can deliver higher fault currents (e.g., utility source and parallel-connected strings of modules) into that circuit than the current-limited source, shall be protected with an overcurrent protective device at the higher current source end of the circuit [690.9(A)].

An Exception to 690.9(A) also exempts the overcurrent device requirement on PV source circuits or dc-to-dc converter outputs when the available short-circuit currents from all sources do not exceed the maximum overcurrent device current rating for a specific PV module or dc-to-dc converter.

Another Exception to 690.9(A) also exempts PV source circuits from overcurrent protection where there are no external sources of overcurrents from parallel connected source circuits or backfeed from inverters (and probably charge controllers). The backfeed capability of inverters under array fault conditions should be a specification in the inverter instruction manual. This specification may not be clearly stated in the instruction manual and a change to UL Standard 1741 may be required to clarify the requirement and define an appropriate test.

Ratings. Most overcurrent devices used in PV systems will be rated at 125 percent of the maximum circuit current determined in Section 690.8 and where used in dc PV circuits, shall be listed for the PV application (photo 6.11). The overcurrent devices listed for the PV application will usually have a greater operating temperature range, from -40°C to 50°C (versus 40°C for a standard OCPD), and have different time history curves.

In very rare cases, an overcurrent device in an assembly may be listed for continuous operation at 100 percent of its rating [690.9(B)]. These assemblies are vanishingly rare in PV systems. This OCPD/assembly 100% combination rating should not be confused with the fact that UL Standards require that all overcurrent devices hold at 100 percent rating and blow/trip within one hour at 135 percent of rating. These tests are conducted on the bare (well-ventilated) OCPD device without an enclosure. A load center, for example, is an assembly/enclosure and the circuit breakers installed in it are derated by the 0.8 factor (rating equals 125 percent maximum current) due to internal heating provided by the circuit breakers themselves and the internal conductors.

Undefined Device. The adjustable electronic overcurrent devices addressed in 690.9(B)(3), 690.8(B), and 690.8(B)(3) appear to be adjustable trip circuit breakers, and after setting per 240.6, they should be applied as any other circuit breaker. The author strongly feels that because they have no current-reducing function, other than the normal trip setting, they should *not* be used to establish conductor ampacity as apparently allowed by 690.8(B) and 690.8(B)(3).

> "Circuit conductors shall be sized to carry not less than the larger of 690.8(B)(1) or (B)(2) or where protected by a listed adjustable electronic overcurrent protective device in accordance 690.9(B)(3), not less than the current in 690.8(B)(3)."

Section 690.8(B) appears to imply that these

devices can reduce a high current to a lower current and then a smaller conductor associated with the lower current can be used when protected with this type of OCPD. As far as can be determined, there are no devices that can provide power circuit (branch circuit) overcurrent limitation/protection in this manner outside of some very sophisticated electronic devices. Hopefully, the 2020 *NEC* will clarify this issue. The 2020 *NEC* may also define how actual current-limited inputs on some devices and energy management systems will deal with limiting the current from multiple sources connected to a single device or circuit. Look to Article 705 and 750 for changes in this area.

Optional Overcurrent Protection. Under the 2017 *NEC*, other than very small solidly grounded PV arrays (690.41), most of the dc circuits in PV source and PV output circuits will *not* be connected directly to earth (solidly grounded). In previous editions of the *NEC*, these ungrounded PV arrays would require an overcurrent device in both (positive and negative) of the ungrounded circuit conductors [2014 *NEC*, 690.9(E)]. Section 690.9(C) in the 2017 *NEC*, now permits the use of an overcurrent device (where overcurrent protection is required) in only one of these conductors, not both. If that option is selected, all dc overcurrent devices in other PV source and output circuits must be in the same polarity conductor in those circuits.

Fusing of DC PV Module Circuits in Utility-Interactive PV Systems

In most electrical systems, the *NEC* requires every ungrounded circuit conductor to be protected from overcurrents that might damage that conductor. Overcurrent protective devices (OCPDs), either fuses or circuit breakers, provide that function. However, some of the smaller utility-interactive PV systems may not need OCPDs in the dc circuits that are connected to the PV modules.

The *NEC* assumes that each ungrounded conductor is connected to some source of overcurrents that might potentially damage that conductor under fault conditions. This source could be a power supply, a utility service, or a battery that supplies currents in excess of the ampacity rating of the conductor. The *NEC*, in 240.21, requires that the conductors be protected at their source of supply. Photovoltaic modules are current-limited devices, and their worst-case, continuous outputs for *Code* calculations are 1.25 times the rated short-circuit current. Therefore, the module cannot generate sufficient current to damage the conductor attached to it in a short-circuit condition. An exception to Section 690.9(A) allows conductors and PV modules to be used without OCPDs where there are no sources of external currents that might damage that conductor or PV module.

Additionally, UL Standard 1703 requires that modules must have an external series OCPD if *external* sources of current can damage the *internal* module conductors. The module can be damaged if reverse currents are forced through the module (due to an external or internal fault) that are in excess of the values of the maximum series fuse marked on the label on the back of the module. Again, if there are no sources of external currents that exceed this marked value, then no OCPD is needed to protect the internal module wiring.

External sources of current (apart from the module or series-connected strings of modules) vary from system to system. These currents can originate from modules or series-connected strings of modules that are connected in parallel to the module of interest, from batteries in the system, or from utility currents backfeeding through utility-interactive inverters.

In systems with batteries and charge controllers, the batteries are a predominate source of currents and OCPDs will be required on each module or series-connected string of modules.

Photo 6.12 • DC source circuit (string) combiner at top.

Photo 6.13 • DC source circuit combiner with output disconnect. Note protective "dead front" shield over exposed energized surfaces/terminals.

Usually, only one OCPD will be required to protect all modules connected in a single series string. A properly sized and located OCPD will protect not only the conductors, but also the modules from external overcurrents [690.9(E)].

The ratings of the overcurrent devices in the combiners must be consistent with the ampacity of the conductors connecting the modules in the source circuit and not more than the rating of maximum series fuse marked on the back of the module. The combiner might be viewed as a branch-circuit load center connected in reverse acting like a PV source-circuit combining panel. The overcurrent devices in this enclosure are located in the proper place in the PV circuits to meet *NEC* and UL requirements. These combiners should not be mounted in direct sunlight to minimize the internal temperature rise (photos 6.12 and 6.13). Inspectors should check for screw-cover enclosures and warning labels if the combiner contains circuit breakers or fuses and has internal, exposed, energized terminals. UL Standard 1741 may be changed to reflect the requirement for such warning labels and, in the future, may require the combiners to be dead front (no exposed energized surfaces) when opened. Some combiners are presently dead front.

In utility-interactive systems, a few inverter designs may be capable of allowing current from the utility to flow backwards through the inverter into faults in the PV array. Systems using these types of inverters would typically require an OCPD at the inverter dc inputs or an OCPD on each string of modules or OCPDs in both locations (see Figure 6.1).

Many of the smaller utility-interactive inverters (below about 6 kW) are designed so that they cannot backfeed currents from the utility into array faults. However, there are currently no specific tests in UL 1741 to validate the lack of backfeeding from the utility, so a manufacturer's certification should be obtained that the inverter cannot backfeed from the

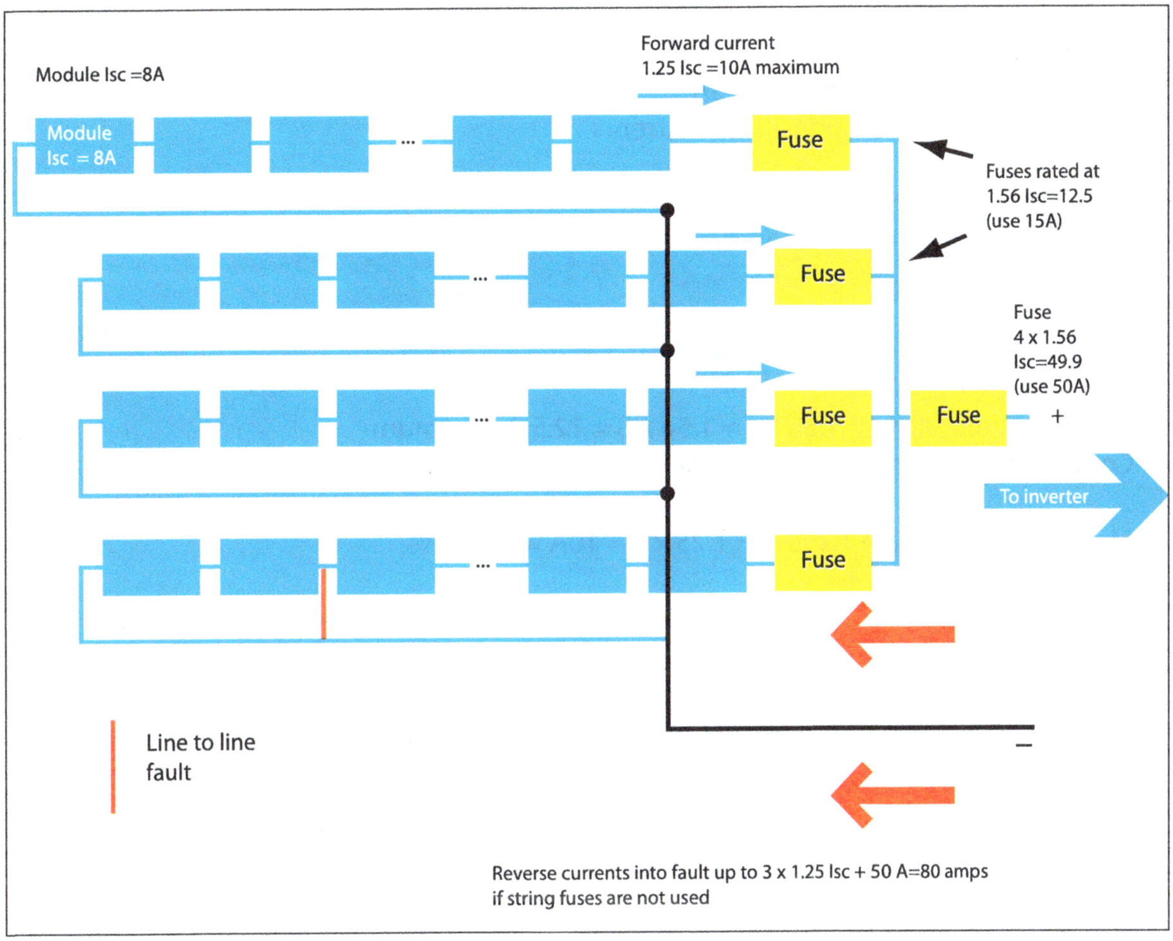

Figure 6.1 • Part of a large system. Possible reverse currents from inverter and parallel strings into fault in one string. Note: all fuses are in the positive conductor.

utility into an array dc wiring fault.

String fuses are required, and an inverter fuse is required if the inverter can backfeed into fault in dc system. The 2017 *NEC* permits the use of only one fuse in each circuit, but any dc fuses in other circuits must be in the same polarity conductor.

The General Case—For Most Larger PV Systems

The most common situation occurs in systems where there are multiple strings of modules connected in parallel. The non-faulted strings may be able to supply sufficient overcurrents (through the parallel connection) to damage either the conductors or the modules in the faulted strings. A basic question is: How many PV modules or strings of modules can be connected in parallel and still meet the *NEC* and UL requirements (marked on the back of each module) before an OCPD is needed on each module/string of modules? UL marks the modules based on reverse-current tests. The *NEC* requires that the manufacturer's instructions and labels be followed [110.3(B)]. The intent of the module marking is to protect the conductors internal to the module at the marked level from reverse currents. This is a maximum value for the OCPD. Lesser values can be used if they meet the *NEC* requirement of 1.56 x I_{SC} to protect the conductor associated with the module or string of modules. In some cases, the value of the module protective overcurrent device is

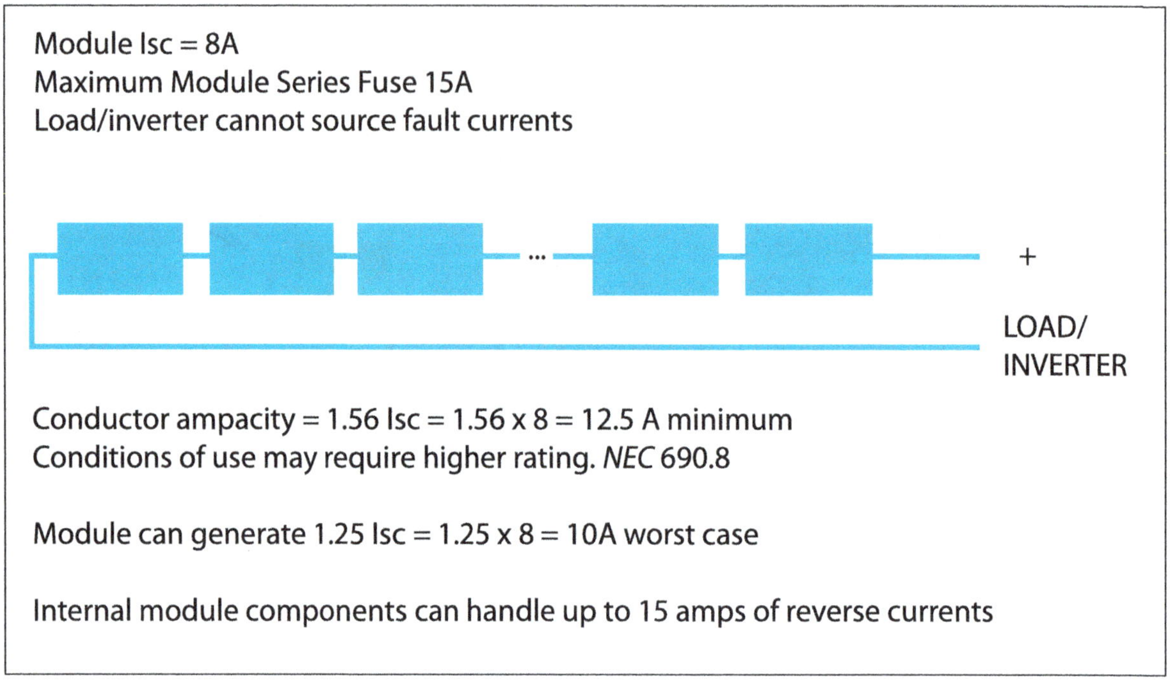

Figure 6.2 • Single string of PV modules. Where there are no external source of overcurrents that can damage the modules or the conductors, no overcurrent protection is required. *NEC* 690.9(A) Exception.

less than 1.56 x I_{SC}. This poses a *Code* conflict [110.3(B) vs. 690.8/9] and is an issue for UL or the listing agency to rectify.

Many installers of 12-, 24-, and 48-volt PV systems ignore the module OCPD requirement and connect modules/strings in parallel. Can it be done and how? David King, when he worked at Sandia National Laboratories, and the author have smoked (destroyed) a few modules and determined that the module OCPD requirement is valid, even at low voltages.

It is easy to see that in a one-string system, an OCPD is needed only when the inverter or battery is a source of overcurrents. No fusing would be required in a one-string system if there were no battery or inverter that could source overcurrents (see Figure 6.2).

Consider *n* modules or strings of modules connected in parallel. The *NEC* requires that an OCPD be installed in the combined paralleled output of all strings (modules) to protect the cable from reverse currents from batteries and backfeed of ac currents through an inverter. In this case, we are assuming that the inverter or the batteries are a potential source of overcurrents. The OCPD will have a minimum rating of 1.56 x n x I_{SC} amps. It is sized at this value to allow maximum forward currents from the array to pass through without interruption and to keep the overcurrent device from operating at more than 80 percent of rating.

Examine a circuit where there are ***n*** modules/strings connected in parallel. Place a ground-fault in one module/string. Examine the sources of fault current that would affect that module string. Let us ignore current from the faulted module/string itself since the wiring in that string is already sized to carry all forward currents generated in the string.

First, there is the potential back feed current from the battery or the inverter in those systems with these components. It is limited to the *NEC*-required OCPD of 1.56 x *n* x I_{SC} between its input and the combined array output. This current is added to the current from the remaining modules connected in parallel. In this case, the current is (n-1) x 1.25 x I_{SC}. The 1.25 is required [representing the maximum current defined in (690.8)] because of daily-expected irradiance values that are greater than the STC-rated I_{SC}.

I-fault = 1.56 x n x I_{SC} + (n-1) x 1.25 x I_{SC}

With a little algebra, the resulting fault current is:

I-fault = (2.81 x n-1.25) x I_{SC} amps. (Fault Current Equation.)

Note that this equation does not account for rating roundup of the OCPD, so each system must be checked with the actual OCPD values.

If the module can pass the UL reverse current test at this I-fault value or greater and be so marked (the maximum protective series fuse on the label), then it is possible to parallel n modules/strings (pick your n) without a series OCPD for each module/string.

For example, a PV module is rated at 60 watts and has a maximum series OCPD requirement of 20 amps, which is marked on the back of the module. The I_{SC} for this module is 3.8 amps. Here are the required calculations and checks for two strings in parallel.

The paralleled circuit OCPD installed at the output of the two paralleled strings will be 2 x 1.56 x 3.8 = 11.8 amps. Assume a 12-amp OCPD is used because the *NEC* now requires module/string OCPDs in one-amp increments up to 15 amps. Fuses are available in these values except there is a jump from 10 to 12 and then to 15. This OCPD could potentially allow 12 amps of fault current to reach the faulted module/string from backfeed from a charge controller/battery or from the utility grid through a utility-interactive inverter.

Another 1.25 x 3.8 = 4.75 amps could come from the parallel-connected module/string for a total of 16.75 amps. This is acceptable because this module is marked for 20 amps.

However, if we try to parallel three of these modules/strings, the fault current equation yields a fault current of 29+ amps that exceeds the 20-amp limit on the module. The single OCPD is 3 x 1.56 x 3.8 = 17.8 amps (because OCPDs at this rating are not common, a 20-amp OCPD must be used). The two parallel-connected modules contribute 2 x 1.25 x 3.8 = 9.5 amps for a total potential fault current of 29.5 amps. This is significantly above the maximum series protective fuse of 20 amps.

In most cases, it is not possible to parallel more than two modules/strings with a single OCPD unless the marked maximum series OCPD is very large in relation to I_{SC} for the module. Some of the thin-film technologies may be able to do this and that will be an installation benefit for them.

Questions about driving voltages to produce these currents? The faults can occur anywhere in the module/string, so a fault involving a single cell could be the trouble spot, and driving voltages over one volt could produce the reverse currents.

What about currents generated within the faulted module string? In the portion of the module/string below the fault (toward the grounded end of the module/string), the currents flow in the forward direction toward the fault and may or may not cause problems. As far as the contribution to the fault current is concerned, the contribution only appears in the fault path/arc and does not affect the ampacity of the conductor. Above the fault (toward the ungrounded end), the currents in that portion of the module/string appear to oppose the external fault currents that are trying to reverse the flow of current, but the string is reversed biased, and the external driving currents are flowing. Because the location of the fault cannot be controlled ahead of time, worst-case currents must be assumed.

The increased marking value of 20 amps on the example module allows for two modules/strings to be connected in parallel and it does make it easier for the installer to use a single OCPD with

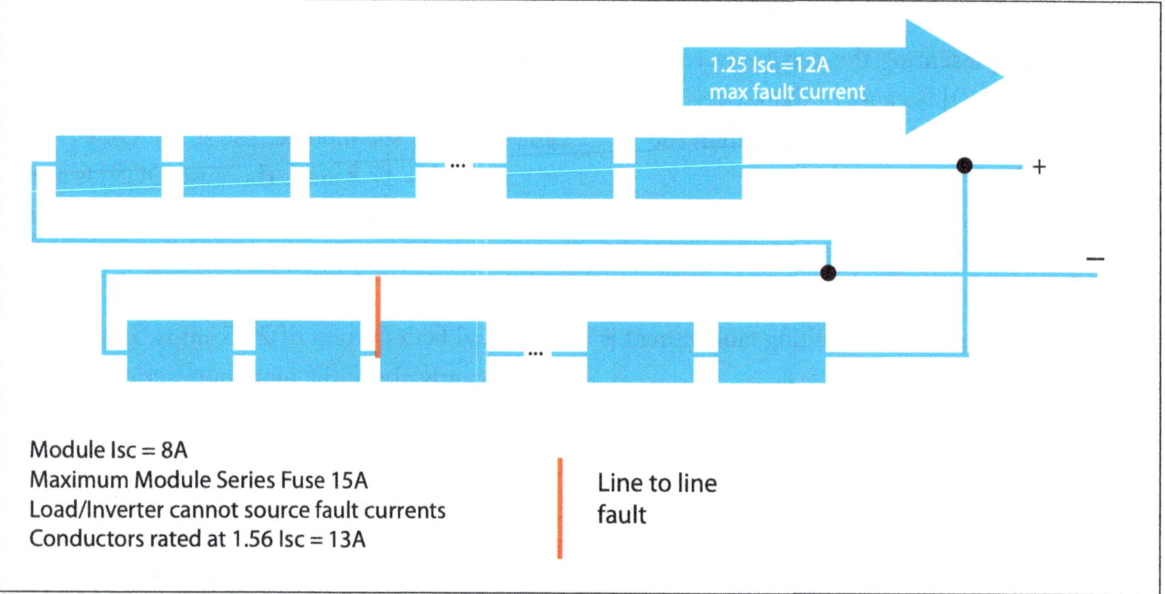

Figure 6.3 • Two strings in parallel. No source of overcurrents that can damage modules or conductors in faulted string. No backfeed from inverter. No overcurrent devices required. *NEC* 690.9(A) EX.

a larger conductor to meet both the *NEC*-required conductor protection and the UL-required module protection with one large OCPD instead of a two smaller OCPDs plus a larger OCPD.

Conductor ampacity must also be addressed if modules are going to be paralleled on a single OCPD. The conductors for each string must be able, under fault conditions, to carry the current from the other parallel strings (modules) plus the current that may be backfed from the inverter or battery. In the case with n strings in parallel and a single OCPD in the combined output, the conductor ampacities would be as follows:

Each of the string conductors would have to have an ampacity of 1.25 $(n-1) \times I_{SC} + 1.56 \times n \times I_{SC}$. If the equation is factored, the required ampacity becomes $A = (2.81 \times n - 1.25) \times I_{SC}$. As before, OCPD roundup is not considered and the values should be recalculated with actual OCPD values. The combined output-circuit conductors would require an ampacity of $1.56 \times n \times I_{SC}$.

Modern, utility-interactive inverters

Many utility-interactive inverters on the market have redundant internal circuitry that prevents currents from being backfed through the inverter from the utility to faults in the PV array. This removes one source of currents in the above equation. With these products, it is possible to have two (and sometimes more) strings of modules in parallel with no OCPDs in the dc circuits. The inverter manufacturers should be contacted for information in this area. The above equations can be modified by deleting the combined-circuit OCPD and then solved to determine both the requirements for OCPDs and the necessary ampacity of the conductors.

In this case, the reverse current flowing through the forward fuse ($n \times 1.56 \times I_{SC}$) is set equal to zero or removed from the equation. In a system with n strings of modules connected in parallel, if one of the n strings develops a fault, the fault current is now reduced to:

$I_{fault} = (n-1) \times 1.25 \times I_{SC}$. For two strings in parallel, $n = 2$ and the fault current becomes:

$I_{fault} = 1.25\ I_{SC}$.

The *NEC* requires that all PV wiring generally be

Photo 6.14 • Retaining kit for a backfed circuit breaker. Required where a circuit breaker is backfed from a standalone inverter output.

sized at 1.56 x I_{SC}. The required module series protective fuse is nearly always greater than 1.56 x I_{SC}.

Therefore, in a system with two strings of modules connected in parallel, there are no sources of fault current that exceed the ampacity of the conductors or the requirements for a module protective fuse. No dc string or array fuses would be needed. *NEC* Section 690.9(A) Exception applies. (See Figure 6.3.)

If there are more than two strings of modules connected in parallel, then the calculations outlined above will have to be made to ensure that (n-1) x 1.25 x I_{SC} is less than the module series protective fuse value. If not, fuses should be used in each string. Again, the actual value of available fuses should be used in the calculations.

Overcurrent Protection in Related Articles. Article 705 establishes additional requirements for overcurrent protection in the ac inverter output circuits. Article 710 does the same for ac circuits in stand-alone systems. In general, conductors shall be protected from all possible sources of overcurrent (705.30, 705.65, and Article 240).

If circuit breakers are backfed, they shall be suitable for such operation [705.12(B)(4)]. This usually indicates that they will not be marked "Line" and "Load." Also, if a backfed breaker is fed by a listed interactive source (PV utility-interactive inverter), the breaker does not have to be mechanically fastened to the panelboard [705.12(B)(5)]. However, where a backfed breaker is connected to the output of a standalone inverter (a voltage source and not interactive),

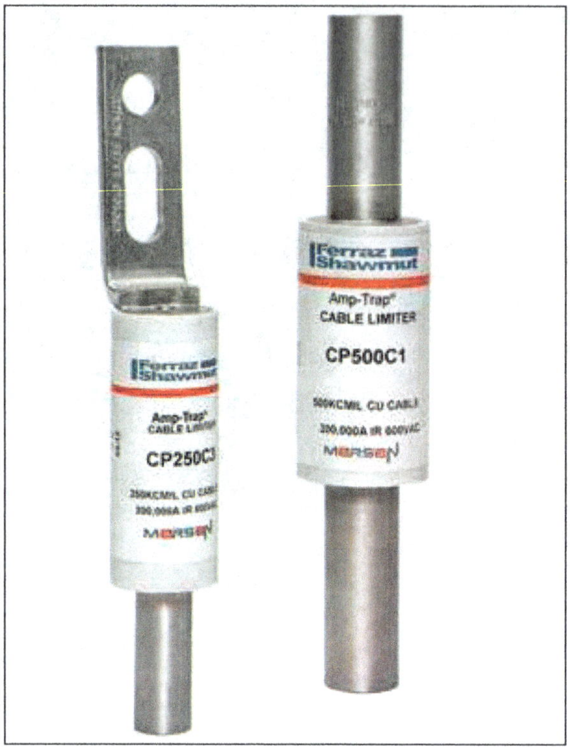

Photo 6.15 • Cable limiters. Attached directly to cables or bolted to busbars. No overload protection is provided. Last-ditch cable protection from bolted line-to-line and ground faults.

that breaker shall be mechanically fastened to the panelboard per 408.36(D) [photo 6.14].

Section 710.15(E) on standalone systems is not clear in this area, because it mentions that backfed circuit breakers connected to *an interconnected* supply (interactive inverter?) require fastening, but the section does not address the output of a standalone (not interconnected) inverter. In Article 705, interconnected sources require or imply the use of interactive sources and those sources, when connected to a backfed breaker, do not require that breaker to be secured.

The interrupt and short-circuit ratings of the equipment in an interconnected system shall consider the contributions from all sources (705.16).

Section 705.31 requires that an overcurrent device be located within 3 m (10 ft) of the connection point of the service conductors in a supply-side connected PV system [705.12(A)]. This overcurrent device would usually be associated with the start of the feeder leading to the output of a utility-interactive inverter. An exception allows this overcurrent protection to be located more than 3 m (10 ft) from the connection point if either a current-limiting breaker or a cable limiter (photo 6.15) are installed at the connection point to the service conductors.

The author notes,that current limiting breakers have different mechanical and electrical characteristics than cable limiters. Where current limiting breakers include a disconnect function and will initiate a feeder to the inverter output at the breaker location, cable limiters have no overcurrent rating (they are sized based on the conductor size) and cannot provide that disconnect function.

While both devices may be current limiting, the cable limited provides only a last-ditch protection before the cable melts under severe overcurrent conditions. The current-limiting circuit breaker provides that protection, possibly at a lower current limit, and provides the overload protection necessary to establish the start of a feeder.

07

Utility Interconnections

Connecting the utility-interactive inverter to the utility grid properly is critical to the safe, long-term, and reliable operation of the entire system. The ac output circuit requirements and the circuits that carry the inverter current in the premises wiring are somewhat complex. However, meeting *Code* requirements can and should be accomplished to ensure a safe and durable system.

Inspectors need to know this material and how to apply it because many PV installers are not familiar with the details of the requirements. The first step is to get the power out of the inverter.

Inverter Output Circuit

In general, the output circuit on an inverter to the first overcurrent device should be sized at 125 percent of the rated output current (the maximum current) of the inverter that, in turn, is determined from the specifications or by dividing the rated power by the nominal ac output voltage [690.8(A)(3)]. For example, a 2500-watt, 240-volt inverter will have a rated output current of:

2500 / 240 = 10.4 amps

1.25 x 10.4 = 13.02

A 15-amp circuit breaker is the next larger standard size and at least 14 AWG conductors should be used. In a few cases, the inverter manufacturer may specify a higher rated current based on a lower than nominal operating voltage (the low end of the anti-islanding range) and this value of rated current should be used. Voltage drop and conditions of use considerations may require a larger conductor. Higher rated overcurrent devices should not normally be used; even though they may be appropriate for the conductor being used, they may not provide required protection for internal inverter circuits. The inverter manual will generally specify the maximum allowable ac output overcurrent device rating and this value should be followed [110.3(B)].

Even though power and current flow from the inverter to the utility, it should be noted that the utility-end of this circuit is where the currents originate that can harm the conductors when faults occur. Utility-interactive inverters cannot generate surge currents and usually shut down under conductor fault conditions. Any overcurrent protection should be located at the utility end of the inverter ac output circuit and not at the inverter end of this circuit [690.9(A)].

In load-side connected systems, it is good practice to install the inverter near the backfed load center so that the backfed breaker commonly used to interconnect the inverter with the utility can also be used as the ac inverter equipment disconnect/isolator required by 690.15. This places the overcurrent device at the utility-supply end of the circuit and groups the ac disconnect for the inverter with any internal or nearby dc disconnect/isolator. A disconnect/isolator must be located within 3 m (10 ft) of the inverter if the inverter is located more than that distance from the backfed circuit breaker [690.15(A)].

Load-Side and Supply-Side Utility Connections

There are two types of connections allowed by the *Code* for interfacing the output of the utility-interactive inverter to the utility power. They are made on either the supply side [705.12(A)] or the load side [705.12(B)] of the main service disconnect of a facility or structure. The load side of the main service disconnect is the most common connection used for the residential system and the smaller commercial system under about 10 kW. Section 705.12(B) covers the requirements and it is heavy reading at best.

Supply-Side Utility Connections

Many larger PV systems cannot meet the requirements/restrictions for a load-side (of the service disconnect) connection to the premises wiring system and a supply-side connection must be considered.

Code Considerations

The supply-side connection (also known as a service-entrance connection) is allowed by the *NEC* and is addressed in several sections in the *Code*.

Section 705.12(A) allows a supply- (utility) side connection as permitted in 230.82(6). Section 230.82(6) indicates that solar photovoltaic equipment is permitted to be connected to the supply side of the service disconnect.

It is evident that the connection of a utility- interactive PV inverter to the supply side of a service disconnect is *similar* to connecting a second service-entrance disconnect to the existing service. Because these PV conductors are unprotected like service-entrance conductors, many of the rules for service-entrance equipment should be considered. Sections 230.2(A)(5) and 230.82(6) permit these parallel power production systems as additional services. However, they do not meet the definition of a *service* found in Article 100 of the *NEC*.

In general, the "Tap Rules" of Section 240 do not apply unless specifically mentioned because they were not developed to address two sources of power in a tap circuit, nor were they developed to ensure safe operation when one source is an unprotected utility power source.

Although the PV output is not considered a service, the unprotected conductors are exposed to the same potential fault currents as the service-entrance conductors. It is suggested that this PV connection be as robust as any service entrance. Section 230.91 requires that the service overcurrent device be integral with the service disconnect or located adjacent to it. A circuit breaker or a fused disconnect would meet these requirements (photo 7.1). A utility-accessible, visible break, lockable (open) fused disconnect (aka safety switch) used as the new PV service disconnect may also meet utility requirements for an external PV ac disconnect in areas where utilities require such an additional disconnect (photo 7.2). Section 690.13(C) requires that the PV system disconnect connected to the supply side of the service disconnect be listed as suitable for use as service equipment.

Section 230.71 specifies that the service disconnecting means for each set of service-entrance conductors shall be a combination of no more than six switches and sets of circuit breakers mounted in a single enclosure or in a group of enclosures. Section 690.13(D) allows up to six switches or circuit breakers for each PV system.

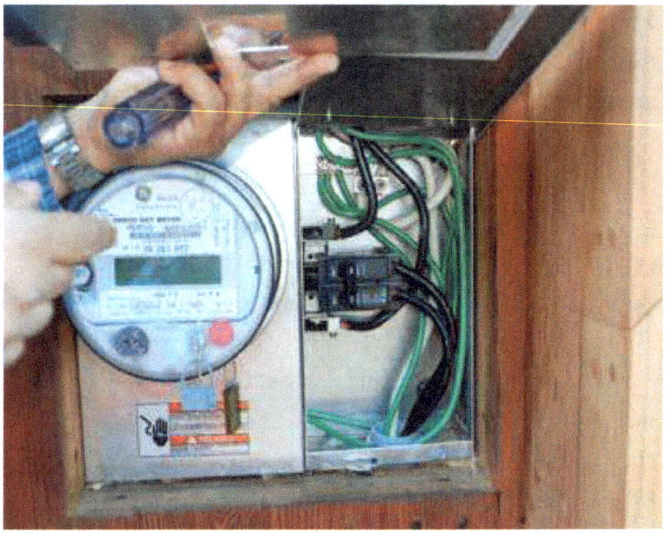

Photo 7.1 • Added lugs provide a supply-side connection. But is it *Code* legal? The added lugs probably violated the listing on the meter/main combo.

Photo 7.2 • Utility-required ac disconnects. They could have been combined with a combiner into one disconnect.

Common industry practice indicates that an ac disconnect associated with a PV system is not counted against the six allowable disconnects for the existing utility service. This is a gray area subject to AHJ interpretation.

Location and Directory

Section 230.70(A) establishes the location requirements for the service disconnect. Sections 705.10 / 690.4 / 690.54 require that a directory be placed at each inverter and service equipment location showing the location of all power sources for a building and the locations of all PV system disconnects (photo 7.3). Locating the PV ac disconnect adjacent to or near the existing service disconnect may facilitate the installation, inspection, and operation of the system (photo 7.4). Many utilities require the service disconnect and the PV ac disconnect to be co-located.

Size Matters

Obviously, the size of the new PV system disconnect on the ac inverter output circuit is important. It will normally be sized at 125 percent of the rated output current from the PV inverter(s). But in small systems, a question arises: How small can it be? Section 230.79 addresses the rating issue for service disconnects. It must be pointed out that these conductors are not service conductors, but for all practical purposes, they are treated as service conductors. Although not a *Code* requirement, many AHJs and PV installers choose to treat these conductors as service conductors. Some inspectors have looked at 230.79(A) and say that it can be as low as 15 amps if that value is at or above the rating of the inverter output circuit. The connection of other specific loads is allowed at this level and is common.

I would suggest caution here, because the connection is to service-entrance conductors rated at 100 amps and above. The typical 15-amp circuit breaker with 10,000 amps of interrupt capability, in this application, may not be able to withstand the available fault current because it is not protected and coordinated with any main breaker. Of course, Section 110.9 should be followed and available fault current calculated. Also,

Photo 7.3 • Directory and labels required by *NEC* and utility.

Photo 7.4 • PV ac disconnect above closed and locked-out/tagged-out service disconnect

a service-entrance-rated 30-amp fused disconnect with 15-amp fuses could be used; listed fuse adapters may be required [see UL Product Spec (formally UL White Book) Category IZZR].

Other considerations are the size of the service-entrance conductors, the new PV connection conductors, and the size of the terminals on available switchgear rated at 30 amps or 60 amps. The added conductors between the existing service-entrance conductors and the new PV system ac disconnect will be subjected to available fault currents and will have no overcurrent protection except that provided by the fuse on the primary of the utility distribution transformer. Making them as large as possible, with an upper limit of the size of the existing service-entrance conductors, would seem prudent, but small disconnects will not accept very large conductors. Cable adapters may be required to make the connection between large service-entrance conductors and smaller terminals on lower current overcurrent devices or disconnects (photo 7.8).

Although these circuits do not meet the definition of service-entrance circuits, the author suggests that Section 230.79(D) be used as the requirement for the smallest PV ac disconnect for PV inverter supply-side connections. Section 230.79(D) requires that the disconnect have a minimum rating of 60 amps. This would apply to a service-entrance rated circuit breaker or fused disconnect. Again, this is not a *Code* requirement, but the suggestion of this text and the author.

Section 230.42 requires that the service-entrance conductors be sized for the sum of the non-continuous loads plus 125 percent of the continuous loads. All currents in a PV system are worst-case and are considered to be continuous. The actual rating should be based on 125 percent of the rated output current for the utility-interactive PV inverter as required by 690.8. The service connection conductors, in this case, must have a 60-amp minimum rating from 230.42(B). Temperature and conduit fill factors must be applied.

For a small PV system, say a 2500-watt, 240-volt inverter requiring a 15-amp circuit and overcurrent protection, these requirements would appear to require a minimum 60-amp rated disconnect with 15-amp fuses; listed fuse adapters would be required. Fifteen-amp conductors could be used between the inverter and the 15-

Photo 7.5 • Meter-main combo. Supply-side connections made by connecting to internal conductors are not allowed. Listing on the combo would be violated.

amp fuses in the disconnect. Section 230.42(B) requires that the conductors between the service tap and the disconnect be rated not less than the rating of the disconnect; in this case, 60 amps.

How we would deal with the 60-amp disconnect, 15-amp overcurrent requirements using circuit breakers is not as straightforward. A circuit breaker rated at 60 amps would serve as a disconnect, and it could be connected in series with a 15-amp circuit breaker to meet the inverter overcurrent device requirements. In this case, the requirements of 705.12(B)(2) should be applied for the series connection between the two circuit breakers.

Sections 110.9, 690.13(F), and 705.16 require that the interrupt capability of the equipment be equal to the available fault current. The interrupt rating of the new disconnect/overcurrent device should at least equal the interrupt rating of the existing service equipment. The utility service should be investigated to ensure that the available fault currents have not been increased above the rating of the existing equipment. Fused disconnects with RK-5 fuses are available with interrupt ratings up to 200,000 amps.

Section 230.43 allows a number of different service-entrance wiring systems. However, considering that the PV connection conductors are unprotected from faults, it is suggested that the conductors be as short as possible with the new PV service disconnect mounted adjacent to the connection point. Section 705.31 allows up to 3 m (10 ft) of unprotected conductors between the point of connection to the service conductors and the first overcurrent protective device.

Making these PV feeder connection conductors as large as the service-entrance conductors, while not a *Code* requirement, would also add a degree of safety. Of course, the added disconnect must be able to accept the larger conductors. Conductors installed in rigid metal conduit would provide the highest level of fault protection.

Neutral-to-ground bonding in the ac PV system disconnect is a gray area. Such a bonding may create parallel paths for neutral currents with two adjacent neutral-to-ground bonds and two grounding-electrode conductors to a common grounding electrode, but illustrations in the *NEC Handbook* in Article 250 indicate that these parallel paths do not cause currents that are objectionable. Utility requirements and the location of the PV production or renewable energy credit (REC) meter should be considered.

The actual location of the supply-side connection will depend on the configuration and location of the existing service-entrance equipment. The following connection locations have been used on various systems throughout the country.

On smaller residential and commercial systems, there is sometimes room in the main load center to connect to the service conductors just before they are connected to the existing service disconnect. In other installations, the meter socket has lugs that are listed for two conductors per lug. Of course, adding a new pull box between the meter socket and the service disconnect is always an option.

Combined meter/service disconnects/load

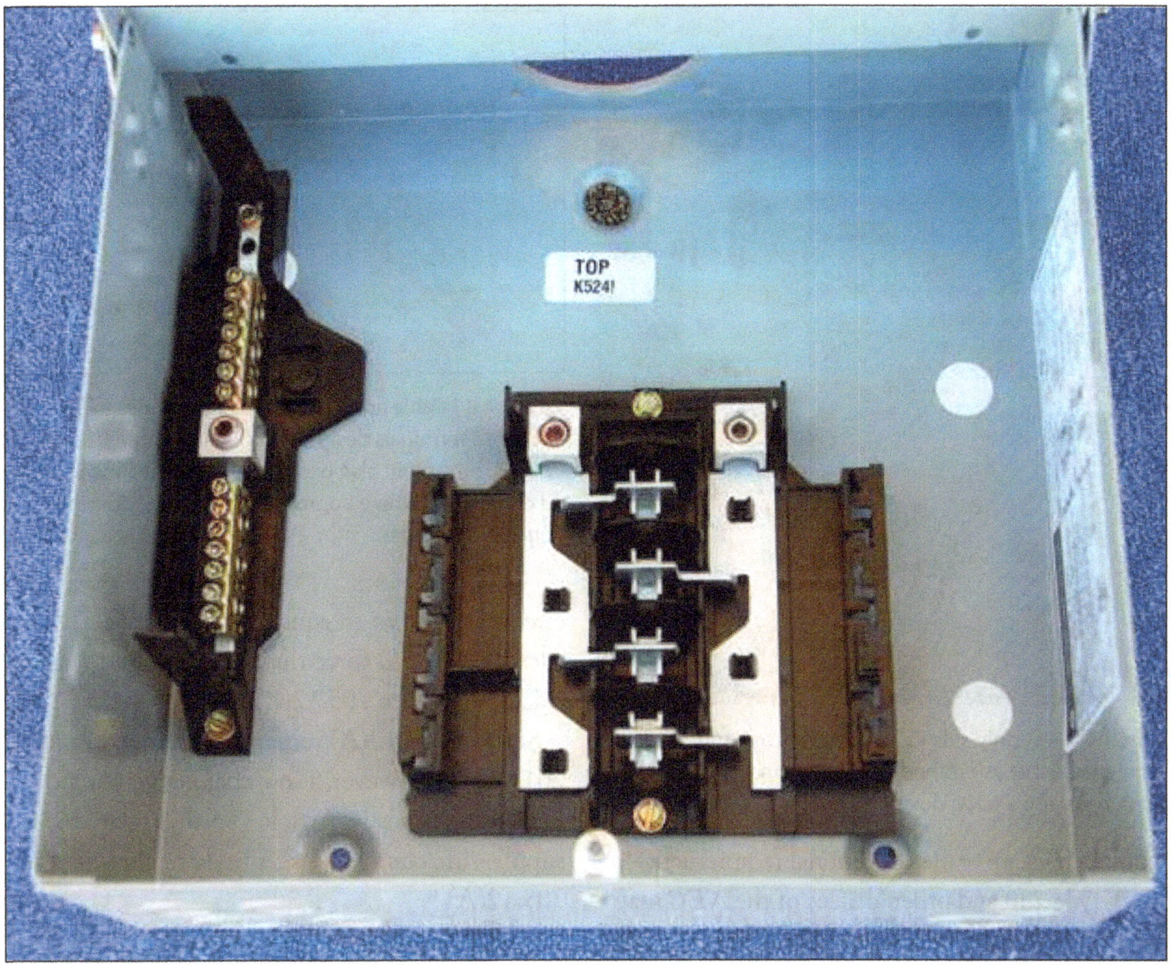

Photo 7.6 • General Electric, 125-amp main-lug-only service-entrance-rated load center. With appropriate breakers, the panel has a 22,000-amp rating. Two stabs can be used to convert the panel to a backfed main breaker panel—but then supply-side connections cannot be made with remaining breaker positions.

centers frequently have significant amounts of interior space where the connection *appears to be possible* between the meter socket and the service disconnect. However, tapping this internal conductor or busbar in a listed device, such as a meter-main combination (aka meter-main combo), would violate the listing on the device and should not be done (photo 7.5).

Where the service-entrance conductors are accessible, a new meter base (socket) could be added ahead of the combination device. A connection box would then be added between the new socket and the combination device. The meter would then be moved from the combo device to the new socket, jumper bars added to the old socket, and the old socket covered.

In larger commercial installations, the main service-entrance equipment will frequently have busbars that have provisions for tap conductors. Any holes intended for connections must be marked "Tap Locations" or similar. The PV connection to this tap point can only be made with the approval and instructions from the manufacturer of the equipment or by the organization supplying the service equipment (usually a UL 508 *Industrial Controls Shop*). These organizations can tap the equipment and maintain the listing on the equipment.

In all cases, safe working practices dictate that the utility service be de-energized before any

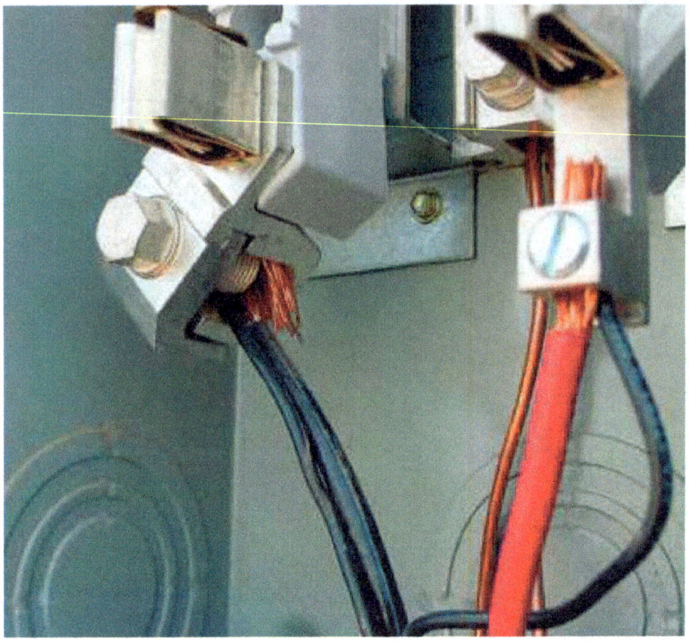

Photo 7.7 • Supply-side PV connection to output terminals of meter socket using conductors that are too small. Double lugging is also a *Code* violation.

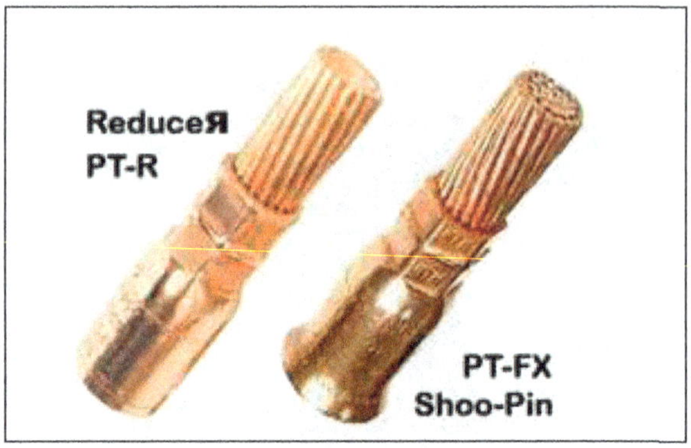

Photo 7.8 • Cable adapters allow larger cables to be connected to terminals designed for smaller cable sizes (reducer). They might also be used to terminate fine-stranded conductors where appropriately rated (shoo-pin). *Courtesy of Greaves.*

connections are made. Additional requirements in Article 230 and other articles of the *NEC* may apply to this PV feeder connection.

In some locations, a six-circuit breaker main-lug-only panel is used as a service-entrance panel. If one of those six circuit breaker positions is unused, it may serve as a supply-side connection for a PV system (photo 7.6). The size of the PV inverter output routed through this circuit breaker could be as large as the service-entrance rating or the busbar rating in the load center, whichever is smaller.

Other Considerations

Although 705.12(A) governs the supply-side connection, as soon as the circuit leaves this new PV service-entrance disconnect, the requirements of 705.12(B) apply if there are any overcurrent devices between the service-entrance overcurrent device and any inverter in the system. Photovoltaic inverter ac combining panels, or load centers needed for multiple inverter installations, and even a circuit breaker in the inverter output circuit serving as an equipment disconnect/isolator (690.15) will force the conductor and busbar calculations into 705.12(B) territory.

Section 705.12(A) remains unchanged in the 2017 *NEC*; however, a few changes were made in other sections in Article 705 that relate to supply-side connections. A basic requirement in 705.12(A) is:

> "The sum of the ratings of all overcurrent devices connected to power production sources shall not exceed the rating of the service."

Note that there is no specific, clearly worded prohibition from having both supply-side and load-side connections on the same PV system if the requirements for each are met. However, caution should be exercised where the sum of the supply-side PV output OCPD and any load-side PV output OCPD exceed the rating of the service. In the absence of loads on the system, PV currents could exceed the service rating.

As an example: A 400-amp service has a 400-amp main panel with a 400-amp main breaker. An 80-amp load-side PV connection is made, and a 350-amp supply-side PV connection is made. Normally loads on the 400-amp panel would absorb some part of the 80-amp

PV output. But, if the loads were not present or turned off, then the sum of the PV currents flowing into the service conductors from both PV systems could total 430 amps, exceeding the service rating.

Sections 705.16, 705.20, and 705.21 establish that all sources and all equipment shall have disconnecting means and that fault currents from all sources shall be considered when determining the required interrupt ratings of those disconnecting means.

Ground Fault Protected Circuits. Section 705.32 requires a PV system supply-side connection where ground fault protection is required for the loads. This would typically occur where a ground fault equipped main circuit breaker was installed as required on solidly grounded wye services at 1000 amps or greater and 150-1000 volts (230.95). An exception allows a load-side connection where the loads are protected from ground fault currents from all sources and that would include ground fault currents sourced by the PV system.

Technical considerations relating to the suitability of GFP main breakers for backfeeding, the selection of an ac ground fault device for a PV inverter output, and the settings of two interacting ground fault devices generally indicate that a PV system should be connected only to the supply side of a GFP main breaker.

One Possible Solution. To avoid the issues associated with those shown in photo 7.7, where space permits, it could be possible to open the service conduit or cable, insert the ac PV fused disconnect in the conduit run and then make the appropriate connections inside the disconnect provided there is sufficient room for such connections and the splicing devices. Using a disconnect with the same rating as the service conductors might facilitate the connections (proper terminal size) as would special multi-conductor terminals (where available) on the line side of the disconnect. Fuse size adapters could be used to match the possibly smaller required PV ac output overcurrent device size to the more robust disconnect fuse holders. This action would probably ensure that the disconnect was service entrance rated and would also minimize (to nearly zero) any additional unprotected conductors. It would also meet 705.31 requirements.

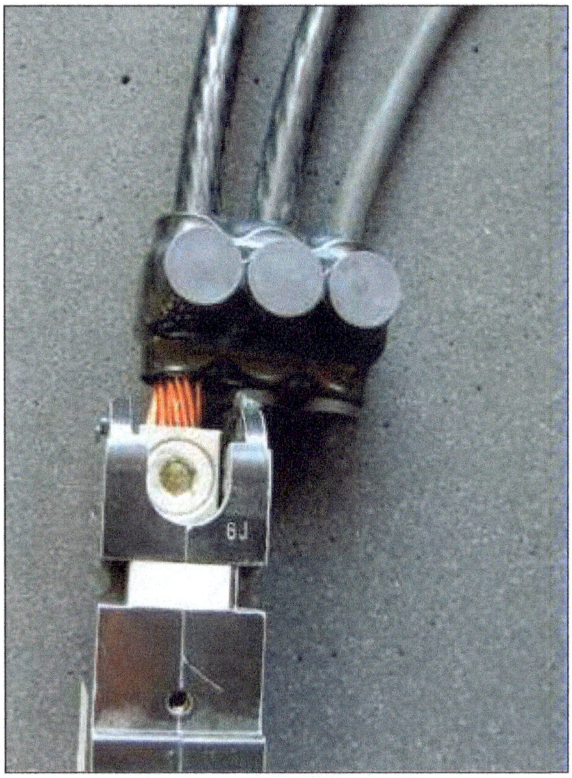

Photo 7.9 • Splicing blocks can be used to tap services with multiple conductors per phase.

Available Fault Currents. Utilities in various parts of the country are making efficiency improvements in their distribution systems. They are installing more efficient distribution transformers in newer subdivisions, and those transformers have lower impedances and lower losses. The result is that the available short-circuit current at the transformer secondary will exceed the commonly used 10,000-amp number, and the service entrance equipment at the building will need to be rated for the higher available fault currents.

The AHJ should require that the PV installer get the specification for the available fault current at the transformer in writing from the utility and provide the calculations for the available fault

Photo 7.10 • There must be an open space somewhere for a backfed breaker for a load-side PV connection.

current at the service entrance equipment. This calculation will also apply to the first disconnect/overcurrent protection used on the supply-side PV connection circuit. In these cases, equipment rated at 22,000 amps or higher may be required. It should be noted that there have been a few cases where the distribution system has been upgraded and the existing service entrance equipment is now underrated.

To provide high levels of protection during faults or line surges (where surge protectors are used) for the PV inverter, it is suggested that fast-acting current-limiting fuses (e.g., type RK-1) be used as the first overcurrent protection device on the PV ac circuit nearest the connection point.

Multiconductor Connections. In some cases, each phase of the service-entrance conductors and the neutral will consist of two or more paralleled conductors. When making the PV connection, all conductors of each phase and the neutral conductor must be "tapped" to ensure that all currents remain balanced [300.3(B)(1), 310.10(H)]. Splicing blocks and other devices are available to facilitate these connections, but they do require space in the enclosures (photo 7.9).

Supply-side service-entrance connections are useful for larger PV systems where the conditions of the load-side connection cannot be met. These supply-side connections normally require that the power be removed from the service to ensure a safe installation. After the first overcurrent device at the PV service disconnect is passed, load-side requirements may be imposed, depending on the complexity of the system.

Load-Side Utility Connections

Code-making panels since 1984 have maintained that 705.12(B) and the prior 690.64(B)(2) will be rigorously applied to any circuits supplied from multiple sources where the circuit is protected by overcurrent protective devices (OCPD) from each source (photo 7.10). Such sources would include the output of PV inverter(s) and the utility supply, and those circuits would include all existing and added premises wiring that could be subjected to currents from either or both sources.

Introductory Paragraph 705.12(B)

> "**705.12(B) Load Side.** The output of an interconnected electric power source shall be permitted to be connected to the load side of the service disconnecting means of the other source(s) at any distribution equipment on the premises. Where distribution equipment, including switchgear, switchboards, or panelboards, is fed simultaneously by a primary source(s) of electricity and one or more other power sources, and where this distribution equipment is capable of supplying multiple branch circuits or feeders, or both,

the interconnecting provisions for the other power sources) shall comply with 705.12(B) (1) through (B)(*5*)."

As in past editions of the *Code*, "distribution equipment" is not specifically defined. Of course, it is usually understood what the term encompasses. But, it also must be understood that distribution equipment in the form of junction boxes for taps or new panelboards can be added to the premises wiring at almost any location that is allowed by the code. So essentially connections for utility-interactive inverters can be made at many points on the load-side circuits that are not in existing distribution equipment.

705.12(B)(1) Dedicated Overcurrent and Disconnect

"Each source interconnection of one or more power sources installed in one system shall be made at a dedicated circuit breaker or fusible disconnecting means."

This section has been clarified. Multiple PV systems may be connected to an existing utility-supplied electrical system, but each connection must be through a dedicated overcurrent protective device and a disconnect. "Dedicated" means that no loads are to be placed on this circuit between the breaker and the PV system ac output.

705.12 (B)(2) Bus or Conductor Ampere Rating

"One hundred twenty-five percent of the inverter output circuit current shall be used in ampacity calculations for the following:"

Note that the title of the section remains bus or conductor ratings and will apply to both as they are defined in the introductory paragraph. This 125% of rated output allowance may slightly reduce the required busbar and conductor ratings required by the following calculations compared to the previous requirement to use the inverter output circuit breaker rating.

705.12(B)(2)(1) Feeder Ampacity

"(1) Feeders. Where the power source output connection is made to a feeder at a location other than the opposite end of the feeder from the primary source overcurrent device, that portion of the feeder on the load side of the power source output connection shall be protected by one of the following:
(a) The feeder ampacity shall be not less than the sum of the primary source overcurrent device and 125 percent of the power source output circuit current.
(b) An overcurrent device on the load side of the power source connection shall be rated not greater than the ampacity of the feeder."

In 705.12(B)(2)(1)(a), a PV connection is made to the feeder somewhere along the feeder, but not at the end opposite the utility connection. The portion of the feeder, from the connection point to the load end of the feeder, can be subjected to currents that are additive and can be as high as the rating of the existing utility end overcurrent device protecting the feeder plus the output of the PV inverter. Hence, the conductor from the connection point to the load end of the feeder must have an increased ampacity equal to the sum of the existing overcurrent device protecting the feeder and 125% of the inverter output rating as noted in this section. (See Figure 7.1.)

In 705.12(B)(2)(1)(b), an allowance is made to protect the existing feeder by installing an overcurrent device on the feeder on the load side of the connection point at the connection point. This allowance, with the added overcurrent protection rated the same as the existing feeder, will allow the existing feeder to be retained and not be replaced as may be required in 705.12(B) (2)(1)(a). The addition of this overcurrent device will prevent excess load currents or faults from exceeding the ampacity of the feeder. The added overcurrent device should be placed adjacent to the power source connection on the load side

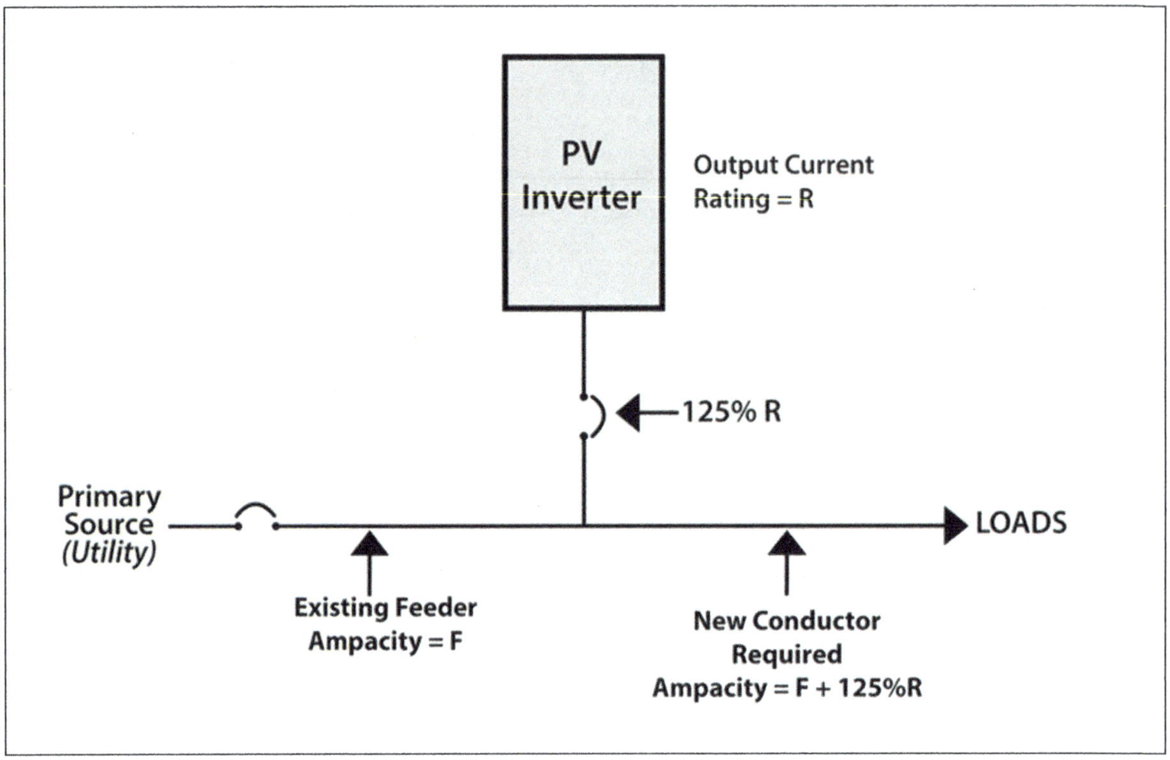

Figure 7.1 • 705.12(B)(2)(1)(a) Increased feeder (on load side of PV connection) ampacity required.

of that connection to protect the remainder of the existing feeder from overcurrents and faults (240.21). Placing the added overcurrent device at the load end of the existing feeder does not provide that protection. See figure 7.2.

This section represents a significant change from past code requirements. It presents requirements for feeder size and overcurrent protection when the utility interactive inverter connection *is not* at the opposite end of the feeder from the utility connection. It should be noted that the ampacity of the feeder between the main OCPD from the utility and the inverter point of connection and the conductor from the point of connection to the inverter output OCPD are not addressed. The following will discuss that omission.

PV Opposite Utility on the Feeder (Not addressed by *Code*—Update Coming in the 2020 *NEC*)

Since the situation where the PV connection *is* at the opposite end of the feeder is not addressed in the new requirements, we can assume (sometimes not a good thing to do) that there is no ampacity correction required on the feeder under that situation. The size of the existing feeder was determined by the existing overcurrent device protecting that circuit from utility currents.

Additionally, while locating the power source output connection at the opposite end of the feeder from the utility source will prevent the feeder from being overloaded by additive currents, it is obvious that 125% of the rated power source output current must not exceed either the rating of the utility-end overcurrent device or the ampacity of the existing feeder.

The general requirement should be (proposed for the 2020 *NEC*): The ampacity of a circuit that has sources protected by overcurrent devices at each end shall be no less than the rating of the largest overcurrent device. This requirement can be used to size feeders or size the rating of the inverter ac output OCPD.

In all cases, the largest OCPD in this situation

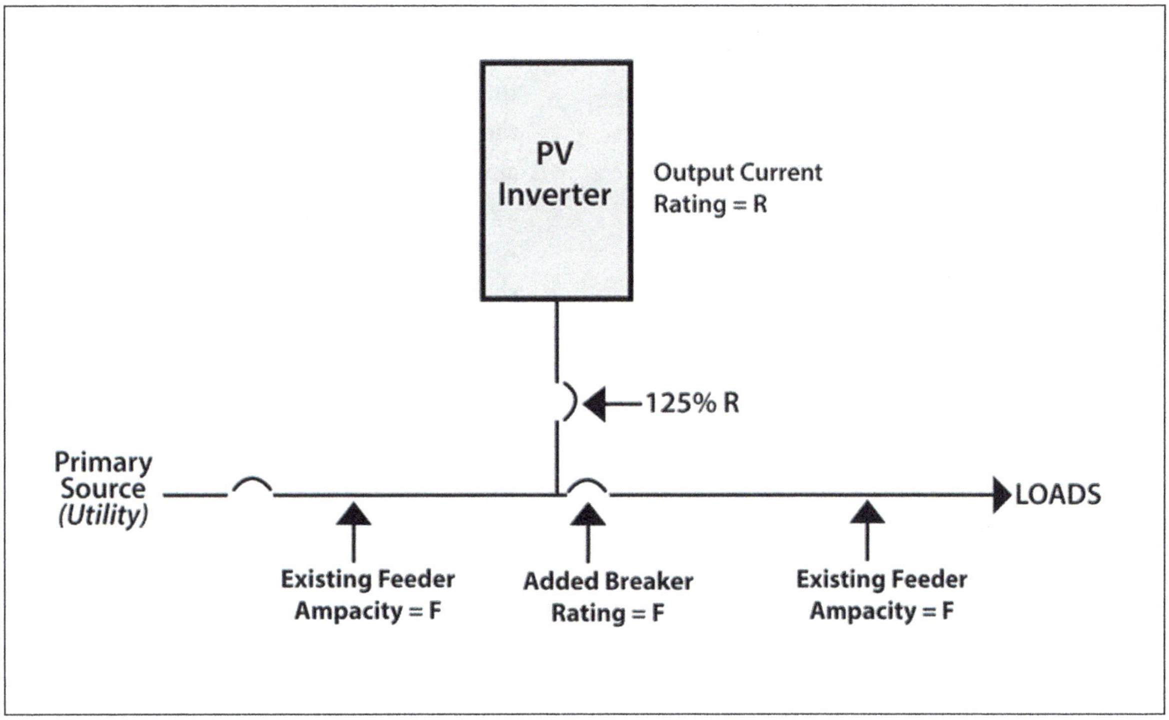

Figure 7.2 • 705.12(B)(2)(1)(b) Additional breaker (on the load side of the PV connection at that connection) required to protect existing feeder from excess currents from PV plus utility.

should be the main OCPD protecting the feeder. Also, where the feeder is not "tapped" at the end point, the conductor from the inverter connection point to the inverter output OCPD should also meet this rule and be sized the same as the existing feeder.

Inverter Output Circuit (the tap conductor) Size

> "**(2) Taps.** In systems where power sources output connections are made at feeders, any taps shall be sized based on the sum of 125 percent of the power source(s) output circuit current and the rating of the overcurrent device protecting the feeder conductors as calculated in 240.21(B)."

This section applies to both existing load taps on a feeder and any taps added after a power source connection has been made to that feeder. All taps, both old and new, will be subjected to this requirement and because the currents used in the various tap rule calculations will be increased due to adding the main OCPD to 125% of the power source output current, in general any tap conductors may have to be increased on existing taps and calculated to be a larger size than normal on new taps.

Busbars

> "**(3) Busbars.** One of the methods that follows shall be used to determine the ratings of busbars in panelboards.

Busbar Rule (a)

> (a) The sum of 125 percent of the power source(s) output circuit current and the rating of the overcurrent device protecting the busbar shall not exceed the ampacity of the busbar.
> Informational Note: This general rule assumes no limitation in the number of the loads or sources applied to busbars or their locations."

This worst-case requirement presented in 705.12(B)(2)(3)(a) assumes that the utility current through the existing main breaker and the current from the output of the utility-interactive inverter may add and that current may create an overload on the busbar. There are no restrictions on the location of the main utility breaker or the PV backfed breaker. If the busbar has a rating equal to the sum of these two values, then no overload would be possible.

It should be noted that reductions in the size of the utility breaker are not prohibited in this section and could be accomplished if allowed by other sections of the *Code*, load calculations and equipment limitations.

Busbar Rule (b)

"(b) Where two sources, one a primary power source and the other another power source, are located at opposite ends of a busbar that contains loads, the sum of 125 percent of the power sources(s) output circuit current and the rating of the overcurrent device protecting the busbar shall not exceed 120 percent of the ampacity of the busbar. The busbar shall be sized for the loads connected in accordance with Article 220. A permanent warning label shall be applied to the distribution equipment adjacent to the back-fed breaker from the power source that displays the following or equivalent wording:

WARNING:

POWER SOURCE OUTPUT CONNECTION–
DO NOT RELOCATE THIS OVERCURRENT DEVICE.

The warning sign(s) or label (s) shall comply with 110.21(B)."

Photo 7.11 • Oops, the electrician got it wrong. The three backfed PV breakers should have been mounted at the top of the panel as far as possible from the main breaker located at the bottom.

705.12(B)(2)(3)(b) is similar to the requirement found in previous editions of the *Code*. If the two sources (utility and PV) are at opposite ends of the busbar, then the sum of those two sources can be as high as 120% of the busbar rating (photo 7.11). With this location of sources, it is not possible to overload the busbar. Note that the busbar must be sized for the loads that are connected. The reason for the value of 120% is lost in history, but may be related to potential thermal overloading of the panelboard. The warning label is self-explanatory and the new requirement referring to section 110.21(B) gives additional information on the specifics of the appearance and durability of the warning label.

Busbar Rule (c)

"(c) The sum of the ampere ratings of all overcurrent devices on panelboards, both load and supply devices, excluding the rating of the overcurrent device protecting the busbar, shall not exceed the ampacity of the busbar. The rating of the overcurrent device protecting the busbar shall not exceed the rating of the busbar. Permanent warning labels shall be applied to distribution equipment that displays the following or equivalent wording:

WARNING:

THIS EQUIPMENT FED BY MULTIPLE SOURCES. TOTAL RATING OF ALL OVERCURRENT DEVICES, EXCLUDING MAIN SUPPLY OVERCURRENT DEVICE, SHALL NOT EXCEED AMPACITY OF BUSBAR.

The warning sign(s) or label(s) shall comply with 110.21(B)."

705.12(B)(2)(3)(c) provides an alternate method of sizing the PV backfed breaker, or determining the size of the required busbar if the PV backfed breaker rating is known. This section will most likely be used when connecting a PV inverter output to a lightly loaded subpanel or when sizing inverter ac combining panelboards. After excluding the main breaker from the utility, the sum of all remaining breakers, both load breakers and the PV supply breaker may not exceed the rating of the busbar. There are several aspects to this requirement that need close inspection and consideration.

First, the main breaker before the addition of any PV has been sized to protect the busbar from possible overload from utility currents. The main breaker will always be equal to or smaller than the busbar rating. For example, many load centers have a 125-amp busbar, but only a 100-amp main breaker. In a normal panelboard or load center, the ratings of the load breakers will total more than the rating of the main breaker or the busbar in nearly all circumstances. If this situation exists, then no PV can be added because the requirement cannot be met because the sum of the load breakers already exceeds the rating of the busbar. However, if the sum of the load breakers were equal to the rating of the busbar, that busbar would still be protected both by the main breaker and by the fact that excess current over the busbar could not be drawn through the load breakers. And again, under this condition, no PV backfed breaker could be added. However, as the sum of the load breakers is reduced, there becomes an allowance for adding a backfed PV breaker with increasing ratings. In the extreme case, there could be a situation where there are no load breakers and only a single or multiple backfed PV breakers rated, in total the same as the busbar. In any of these cases, no matter where the PV breaker is installed on the busbar, the supply and/or load currents cannot exceed the rating of the busbar.

But, it should be noted that in existing load centers, with the sum of the load breakers totaling more than the busbar rating, it is unlikely that load and load breakers can or will be removed.

It would not be wise to install a backfed PV breaker that was larger than the main breaker in those instances where the busbar rating is larger than the main breaker. If this were done, the main breaker could trip from overcurrents through the larger PV breaker.

However, this section needs to be used with caution because there is no restriction on the position of the backfed PV breaker. Suppose a 50-amp PV breaker were installed near the top of the 100-amp busbar in the load center near a 100-amp main breaker and there were 50 amps of load breakers. The *Code* requirement is met with this configuration. However, what happens if someone disregards the warning label or the label simply falls off over time? Added load breakers in the empty positions could pose an overload situation on the busbar. I suspect that many jurisdictions are going to have to emphasize the permanent nature of that warning label to cover the materials that it is made of and the manner in which it is fixed to the panel board. Also, some consideration might be made to permanently covering unused panelboard breaker positions. It might be wise to adopt a local jurisdiction requirement that the backfed PV breaker always be installed as far as possible from the main utility breaker and an additional warning label as required in (b) be placed adjacent to this PV breaker, or other PV overcurrent device.

Center-fed Panelboards and Multiple-ampacity Busbars

> "(d) A connection at either end, but not both ends, of a centerfed panelboard in dwellings shall be permitted where the sum of 125 percent of the power source(s) output circuit current and the rating of the overcurrent device protecting the busbar does not exceed 120 percent of the current rating of the busbar."
>
> "(e) Connections shall be permitted on multiple-ampacity busbars where designed under engineering supervision that includes available fault current and busbar load calculations."

There was no provision in earlier codes to address center-fed panelboards or multiple-ampacity busbars and it was not possible to install the PV breaker at the opposite end of the busbar from the main breaker because there were two or more busbars connected to the main breaker. 705.12(B)(2)(3)(d) was specifically added to the 2017 *Code* to address the common situation where PV needs to be connected to a center-fed panelboard. Although not clearly stated, there was no intent to allow center-fed panelboards to be installed under sections (a) through (c) of 705.12(B)(2)(3). PV connections are now allowed on center-fed panelboards and multiple-ampacity busbars under the conditions noted in these sections. Engineering supervision typically indicates that the analysis of the PV connection will be made and stamped by a professional engineer. The load calculations will look not only at the breakers installed on the busbars, but also the loads connected to those breakers, and the possibility of installing additional breakers and loads in unused spaces in the panelboard. Fault studies may involve looking at the electrical time versus current profiles for each of the circuit breakers involved to ensure that all portions of the busbars will be protected under various fault scenarios from currents sourced both from the utility through the main breaker and from the PV system through the backfed PV breaker.

Marking (3), Suitable for Backfeed (4), and Fastening (5)

These sections are largely unchanged from the *2014 NEC* and are self-explanatory.

Examples

1. A dwelling has a 125-amp rated service panel (busbar rating) with a 100-amp main breaker at the top. How large can the backfed 125% of inverter output current be assuming that it can be located at the bottom of the panel? Circuit breaker round up is allowed.

   PV OCPD + main OCPD <= 120% of panel rating

   120% of panel rating = 1.2 x 125 = 150 amps

   PV + 100 <= 150, therefore the 125% of inverter output current can be up to 50 amps

2. Suppose it was 100-amp panel with a 100-amp main breaker, how much 125% of inverter output current could be added?

   PV + 100 <= 1.2 x 100 = 120

   The maximum PV backfed 125% of inverter output current would be 20 amps.

3. A 200-amp main panel with a 200-amp

main breaker would allow up to 40 amps of 125% of inverter output current, which could be any combination of 125% of inverter output currents that added up to 40 amps on either line 1 or line 2 of the 120/240-V panel.

PV + 200 <= 1.2 x 200 = 240
PV <= 240-200 = 40 amps

4. When working the problem from the inverter end, we start with the continuous rated inverter output current. This is usually the rated power divided by the nominal line voltage, unless the inverter specifications list a higher continuous output current (sometimes given at a low-line voltage).

   A 3500-watt, 240-volt inverter has a rated ac output current of 3500/240 = 14.58 amps.

The output circuit must be sized a 125% of 14.58 = 18.2 amps [690.8(A)(3) and (B)(1)]. The next larger overcurrent device would be a 20-amp OCPD and this would be consistent with the use of 12 AWG conductors if there were not any very large corrections applied for conditions of use or voltage drop. This system could be connected to a 200-amp panel or a 100-amp panel providing the backfed 20-amp breaker could be located at the bottom of the panel (utility input/breaker at the top).

There is sometimes a tendency to use that 30-amp breaker and those 10 AWG conductors that happen to be on the truck. While this would pose no problems for conductor ampacity or circuit protection, the inverter specifications may limit the maximum size of the output OCPD and larger values may not be used [110.3(B)].

No bottom breaker position?

From the above equations, if the backfed PV OCPD cannot be located at the bottom of the panel or at the end of the circuit, it is not possible to install the backfed breaker without changing something. That 120% allowance drops to only 100%. Any panel that has a main breaker rated the same as the panel rating in the above equations would not allow *any* OCPD to be added. The 100%-of-the-panel-rating factor (instead of 120%) would equal the rating of the main breaker, and the equation would force the PV breaker rating to be zero. Of course, in some unusual situations, the other allowances in 705.12(B)(2) might be used.

In a few cases, an *NEC* Chapter 2 load analysis might reveal that the service for the dwelling needed to be only 150 amps, but a 200-amp panel was installed with a 200-amp main breaker just to provide extra circuit breaker positions. In this case, a 150-amp main breaker could be substituted for the 200-amp breaker if the panel is listed for interchangeable main breakers. Even without the bottom position being open, 50 amps of PV breaker could be installed.

Systems with multiple inverters

Many residential and small commercial systems use more than one inverter (photo 7.12). If the local utility requires an accessible, visible-blade, lockable disconnect on the ac output of the PV system, then more than one inverter could not be connected directly to the main panel. The two, or more, inverters would have to have their outputs combined in a PV ac inverter combining subpanel (PV ac subpanel) before being routed through the utility disconnect, where required, and then to the main panel. The utility disconnect (which may now be the PV System disconnect) is not normally fused, but some are, depending on the system configuration (sometimes fused in a supply-side connection). The PV ac subpanel rating, the rating of the disconnect, and the ampacity of the conductor to the main panel are also con-

Photo 7.12 • Three inverters with PV ac combining panel to the left. The combining panel is required where the utility will allow only one ac PV disconnect or one REC meter (not shown).

trolled by 705.12(B) requirements and 705.12(B)(2)(3)(c) might be used.

Here is another example.

The dwelling has a 200-amp main service panel with a 200-amp main breaker and there is an empty breaker position (2-poles) at the bottom of the panel. The utility requires an external disconnect switch and it is desired to install a PV system that has a 3500-watt and a 4200-watt inverter. A PV ac panel will be used to combine the outputs of the two inverters and the output of that PV ac panel will be routed through the utility disconnect and then to a single backfed breaker in the main service panel.

The ratings of the output circuits of each inverter are:

3500/240 = 14.58 amps

1.25 x 14.58 = 18.2 amps; use a 20-amp breaker and 12 AWG conductors.

4200/240 = 17.5 amps

1.25 x 17.5 = 21.875 amps; use a 25-amp breaker and 10 AWG conductors.

The 20- and 25-amp breakers are mounted in the bottom of a main-lug-only panel used as a PV ac subpanel. Normally, no loads will be connected to this subpanel. It will be dedicated to the PV system. If no loads are connected to this ac inverter combining panel, the 705.12(B)(2)(3)(c) would allow the breakers to be positioned anywhere on the busbar. The panel/busbar would be rated at least at 40 amps (18.2 + 21.875 = 40.075) but normally, a 60–100-amp subpanel would be used.

The next step is to calculate the backfed breaker that must be placed in the main service panel to handle the combined output of both inverters from the PV ac subpanel and to protect the conductor carrying those combined outputs under fault conditions from high utility currents.

The combined currents from both inverters are:

14.58 + 17.5 = 32.08

The overcurrent device should be 40 amps (1.25 x 32.08= 40.1), and this meets *Code* requirements for a 200-amp panel with a 200-amp main breaker. Note that the allowance of using 125% of the inverter currents allows a 40-amp backfed breaker, whereas if the inverter output breakers had been used, the resulting 45-amp backfed breaker would have not met *Code* requirements.

With overcurrent devices on each end of the circuit between the inverter output ac combining panel and the 40-amp backfed breaker in the main panel, all conductors and equipment in this circuit should be rated for at least 45 amps which is the total of the two inverter output breakers (20 + 25 = 45). This exceeds the rating of the 40-amp backfed breaker in the main panel and is discussed above where there are overcurrent devices on the ends of a circuit.

One Diagram Is Worth a Thousand Words

Many people do better with diagrams than they do with words, so figure 7.3 should be just up their alley. This big picture diagram can be used with many types of utility interactive PV systems. These systems all start with a meter connected to the utility as shown on the left (blue). After that, we may be dealing with an existing service disconnect and the connected existing load center or with a PV supply-side connection, which is similar a second service entrance on the existing premises wiring system. In either case, the *NEC* requirements of Article 230 apply as noted at the bottom of the diagram. In most jurisdictions, the local utility will require a PV disconnect on the ac output of the PV system, and many areas will use a Renewable Energy Credit (REC) meter to measure the PV system output. As shown, one or more single inverters may be connected or even one or more "strings" of microinverters or ac PV modules may be connected to the added combining panel (green blocks). Or, a single inverter could be connected to an existing load center (red blocks). In some cases, multiple inverters might be connected through an ac combining panel and then backfeed an existing load center. Let's start our examination of the requirements at the inverter end of the circuit.

Inverter Output Circuit

All utility-interactive inverters have a rated output current that cannot be exceeded. There are no surge currents in these output circuits and *NEC* 690.8 requires that the circuit and the overcurrent protective device (OCPD) be rated at 125% of that rated output current. When the calculated OCPD value is a nonstandard value, the next standard higher value should be used, but not to exceed the maximum overcurrent value given in the technical specifications for the inverter. Conductor size should be selected so that it can carry the inverter current and is protected by the OCPD rating.

The asterisk (*) by the 690.8 in the diagram indicates that *if* there is an overcurrent device mounted at the inverter, then the requirements of 705.12(B), *and not 690.8*, will apply. Some installers and manufacturers use a circuit breaker or fused disconnect at the inverter to meet the requirements of 690.15 to have an equipment disconnect at the inverter. The inclusion of an overcurrent device at this location may require the output conductors from the inverter to be larger [705.12(B)] than would otherwise be required by 690.8.

After the First Inverter Overcurrent Device

Any conductor or busbar that can have power flowing from more than one source of supply (under normal or fault conditions) such as the utility and a PV inverter, and where the conductor is protected by an overcurrent device on each

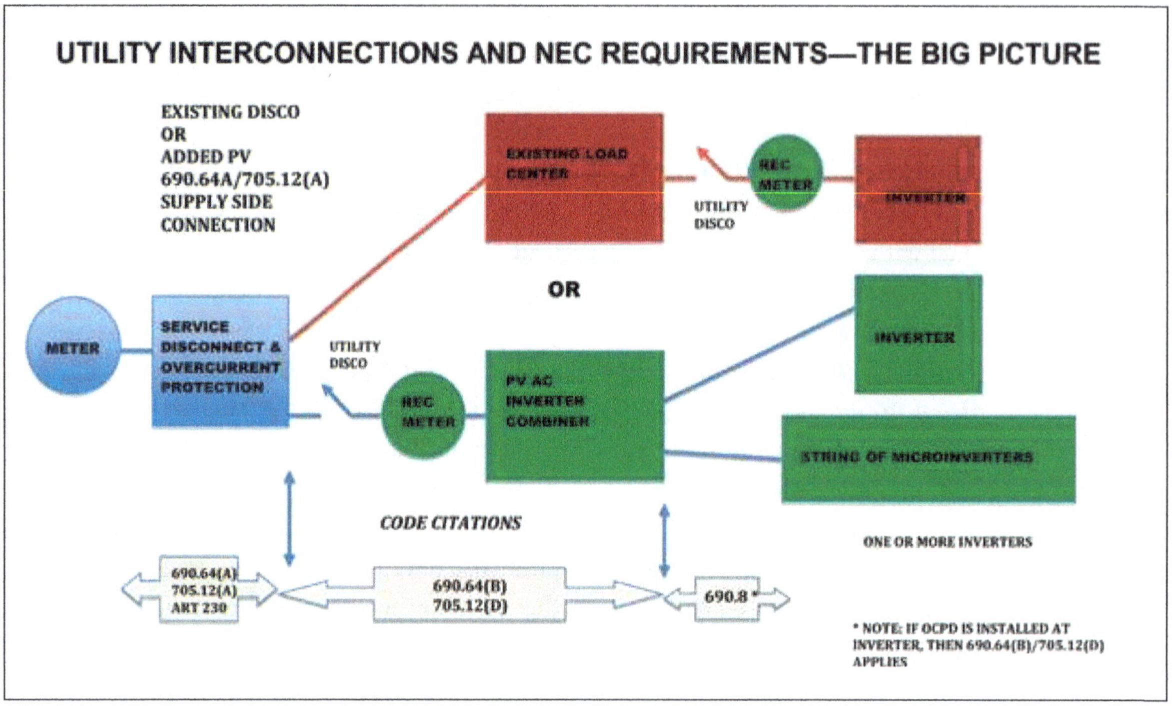

Figure 7.3 • The Big Picture Diagram.

supply source must meet 705.12(B) requirements. This is the longstanding 120% allowance [when 705.12(B)(2) conditions can be met]. Section 705.12(B) is going to apply to all *conductors and busbars* from the first overcurrent device connected to the inverter output all the way to the service disconnect.

These busbars and conductors would include the busbars of any backfed main panel boards connected to one or two inverters or sets of microinverters, and any busbars in PV ac inverter combiner panels. The conductors or feeders between the panelboards or load centers and the main service disconnects are also subjected to the requirements of 705.12(B) as noted on the diagram.

The Main Disconnect and on to the Meter

Any circuit between the meter and the service disconnect would be considered a service-entrance circuit and be governed by the requirements of Article 230. However, after passing through the overcurrent device/disconnect on either an existing service disconnect or through the overcurrent device/disconnect on an added PV supply-side connection, the requirements of 705.12(B) apply all the way to the first overcurrent device connected to the inverter output.

Making the PV Circuit to Service Conductor Connection

Most, but not all, supply-side PV connections are made to the service conductors between the meter location and the main service disconnect. This assumes that the meter is the closest device toward the utility supply and that a net-metering connection is desired, allowed and/or required by the local utility. In some jurisdictions, a *zero sequence* or *cold* meter socket is required by the local utility and the main disconnect may be ahead of the meter socket. In either case, the allowance of *NEC* 705.12(A) is for the connection to be made on the utility side of the main service disconnect.

Local utility requirements will determine where

the PV connection can actually be made, and, of course, the utility will have to be notified to turn off power to the service before the connection is made. It is usually acceptable to open the raceway between the disconnect and the meter, insert a j-box and make the connection using appropriate splicing devices.

As mentioned previously, the PV inverter output connection cannot be made to the internal bus bars or conductors between the meter socket and the main breaker on a meter-main combo panel or in large switch gear unless those conductors or busbars are factory marked as tap points, the manufacturer's instructions and hardware are available, and the utility approves the connection. Otherwise the listing on the combo panel would be violated.

Special, listed PV connection meter socket adapters are available from several sources, but may or may not be approved for use by a local utility. These automatically provide a set of "PV ready" terminals that are connected to the load side terminals of the meter (photo 7.13).

Several load center manufacturers make "ranch panels", which have a 400-amp input busbar and two, 200-amp breakers connected to that busbar. One of the breakers feeds a normal load center busbar and the second 200-amp breaker is available for a supply-side PV connection.

In all cases, the connection will be to service-entrance conductors and any devices, methods or equipment used must comply with the requirements for such connections in this environment. For example, a short-circuit current rating (SCCR) might apply to any meter socket, busbars, fuse holder, disconnect or terminal block which might carry fault currents.

Battery Backed up Systems. PV systems containing energy storage systems (ESS), usually a battery bank, are becoming more common as utility outages due to natural disasters (hurricanes, snow storms, flooding and tornados) become more common and as battery systems for this application proliferate in the industry [Article 706]. See photo 7.14.

Photo 7.13 • Meter adapter used to provide supply-side PV connection. *Courtsey EATON.*

A utility-interactive multimode inverter is used in these applications and that inverter will have one ac input/output that is connected to the utility, a second ac input/output that is connected to the load circuits and a dc input/output that is connected to a battery bank. The PV array may be connected to the multimode inverter as an "ac coupled" system where a separate utility-interactive inverter converts dc PV energy to ac energy and feeds it to the ac load circuit that is connected to the multimode inverter. The multimode inverter charges the batteries from either the ac utility or the ac output from the PV array.

Or, the system may be "dc coupled" where the PV array is connected to a charge controller that charges the batteries directly. The multimode inverter will discharge the batteries into the load

Photo 7.14 • Typical 10 KW ac output whole-house battery-backed-up PV system. System is ac coupled with additional utility-interactive inverters located outside at the PV array.

circuits or the grid when there is excess energy available or charge the batteries from the grid when necessary. See Chapter 5 for additional details.

In either type of system, it is important to note that the ac utility input on the multimode inverter may, at times, be a *load* on the utility feeder or service since it can pass through utility currents to supply loads under normal operating conditions. That input can also supply utility currents to charge the batteries when excess PV energy is not available.

The amount of potential maximum load on the utility supply must be calculated and used in determining conductor sizes, overcurrent and disconnect device ratings and service entrance ratings. The load may be the entire building load when a "whole house" PV battery backed up electrical system is installed *plus* the load due to maximum battery charging. Or, the load may be some subset of the maximum building load where only a few critical circuits are connected to the battery backed up PV system, but the maximum battery charging load must still be added to this load.

Each type of multimode inverter from the several manufacturers will have software controls available to limit and otherwise control the various pass through ac currents to the connected loads, the ac currents that can be made available for battery charging and the maximum ac currents that can be supplied to or supplied from the utility source. The instruction manual for these relatively complex devices must be carefully reviewed to ensure that the appropriate software settings have been made and that the external circuitry is compatible with these adjustments. In general, these software adjustments *must not* be accessible to unqualified people.

Summary of *NEC* 705.12(B)

It is obvious that the new 705.12(B) requirements are significantly different from those in past years. While many of them make sense from an engineering point of view, the real world faced by inspectors and plan reviewers may be somewhat different where people typically ignore instructions, ignore the code, and ignore warning labels. On the other hand, PV systems have not changed significantly from 2011 to 2017 and the electrical systems they are being connected to have not changed significantly, so these new requirements might also be applied in jurisdictions using earlier editions of the *Code* by accepting alternate methods and materials waivers based on the 2017 *NEC* clarifications.

The load-side connection for the utility-interactive PV inverter is not the easiest subject to understand, but the correct application of these requirements will yield a safer, more durable system. When the requirements of load-side connections become complex and expensive, a supply-side connection may be used.

08

Plan Checking and Inspecting

A formal plan review process and plan reviewers are not found in numerous smaller jurisdictions due to budget constraints and the perceived lack of need for such activities. This is probably an acceptable situation where routine inspections involve residential electrical systems and the smaller commercial systems. Those electrical systems have remained essentially unchanged for decades with the exception, of course, of equipment such as arc-fault circuit interrupters and ground-fault circuit interrupters and the like.

The load center that I installed in 1960 in the home I grew up in is nearly identical to the load center I installed in my retirement home in 2010. Yes, each edition of the *National Electrical Code (NEC)* provides changes in the requirements for residential and commercial electrical installations, but the equipment is very similar from year-to-year and the methods of installing that equipment are very familiar to the electricians and inspectors. The electrical contractors are familiar with the equipment and installation procedures and although there are differences from installation to installation, there are many similarities. Mistakes in the installation occasionally get made and are found by the inspectors, corrected and are not made in future installations, hopefully, by the same contractor.

However, we now have the ubiquitous photovoltaic (PV) power system springing up in numerous neighborhoods on residences and commercial buildings throughout the country. The photovoltaic (PV) industry is dynamic and fast changing. New PV modules are being released monthly and each new module has a different size and different ratings. New inverters and other equipment are also coming to market with significant frequency. Although the equipment is connected with the conductors and uses overcurrent devices that we are familiar with, the outdoor environment and the rapidly changing equipment generally point to the need of some sort of plan review process to ensure that the somewhat unique PV *Code* requirements are being met.

Most inspectors would agree that an unfamiliar installation, a new contractor, and unfamiliar equipment as well as new and slightly different *Code* requirements dictate that a longer-than-usual inspection may be required. In fact, it is difficult when inspecting the typical PV system as installed in the field to fully ascertain that all the *Code*

Photo 8.1 • (*top*) Module labels and wiring not readily accessible

Photo 8.2 • (*left*) Conductor markings not easily accessible in roof-mounted dc source-circuit combiner. Pre-2017 *NEC* "grounded" PV array

requirements have been met. The modules are fastened to the roof and the rating labels cannot be seen so the ratings for open-circuit voltage (Voc), short-circuit current (Isc), and maximum protective overcurrent device cannot be verified (photo 8.1). The wiring is typically behind the modules or in conduit and the markings on the conductors in many cases cannot be verified and, like the module labels, some of the equipment such as combiners is not readily accessible (photo 8.2).

Even without a formal plan review process or dedicated plan reviewers, it would behoove the typical inspector to spend a few minutes in the office with the permit application and the plans, which will be described shortly, and verify that, at least, on paper the proposed PV system meets *Code* requirements.

The Plan Review

Of course, to do a plan review one must have a plan. Many jurisdictions do residential electrical permitting across the counter with no substantial, detailed plans required for the electrical system. A form is filled out, the fee is paid, and the permit is issued. At some point, an inspection is requested, and the inspector makes the inspection

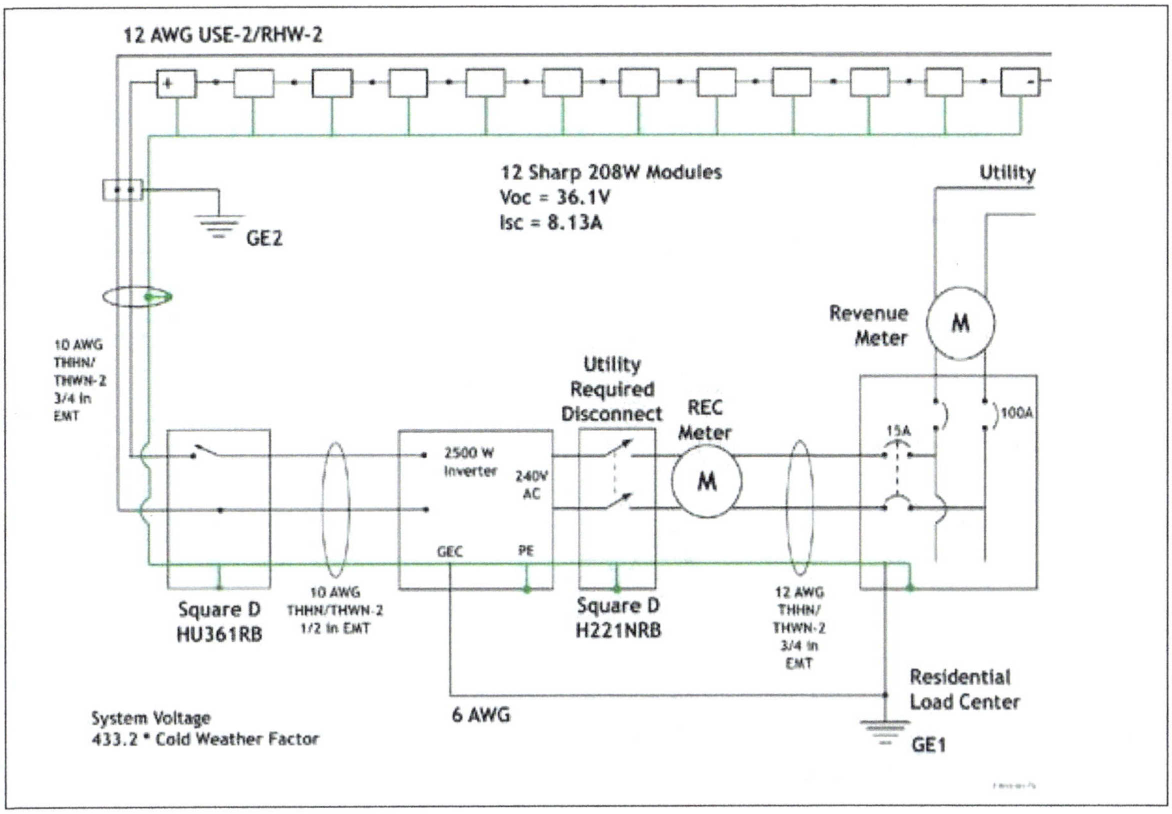

Figure 8.1 • Three-line diagram of a 2.5 KW PV system. PV system disconnect is the 2-pole backfed breaker. Extra 6 AWG GEC from inverter is not required by 2017 *NEC*.

and either approves or disapproves the installation. While this works for normal residential and for some small commercial electrical installations where the equipment and *Code* requirements are fairly well standardized, it is probably not the best option for PV installations, as noted above, where the equipment and the installation techniques are changing frequently.

Some jurisdictions are adopting an expedited PV permitting process for PV systems under 10 kW. This expedited permitting process usually includes a diagram and other information and is, in some cases, based on the process described on the SolarABCs' website. That expedited permitting report and substantial amounts of other PV-related information can be found at: http://www.solarabcs.org/about/publications/reports/expedited-permit/index.html.

The following paragraphs contain some suggestions that have been found useful when performing a plan review to determine if the basic design of the PV system meets minimum code requirements. Figure 8.1 shows a three-line diagram of a small PV system with a single string of modules as an example and such a diagram/schematic should be required in any permit application.

Equipment Lists and Specifications

A list of the equipment used and the specifications for that equipment should be included with the permit. This list would include the PV-specific equipment such as the PV modules, the inverter, the fuses, and circuit breakers. Listing/certification and rating information must be included. The specifications of this equipment are necessary to determine if the conductors have been properly sized and that the fuses and circuit breakers used in the dc parts and ac parts of the system are properly rated. Factory cut sheets or

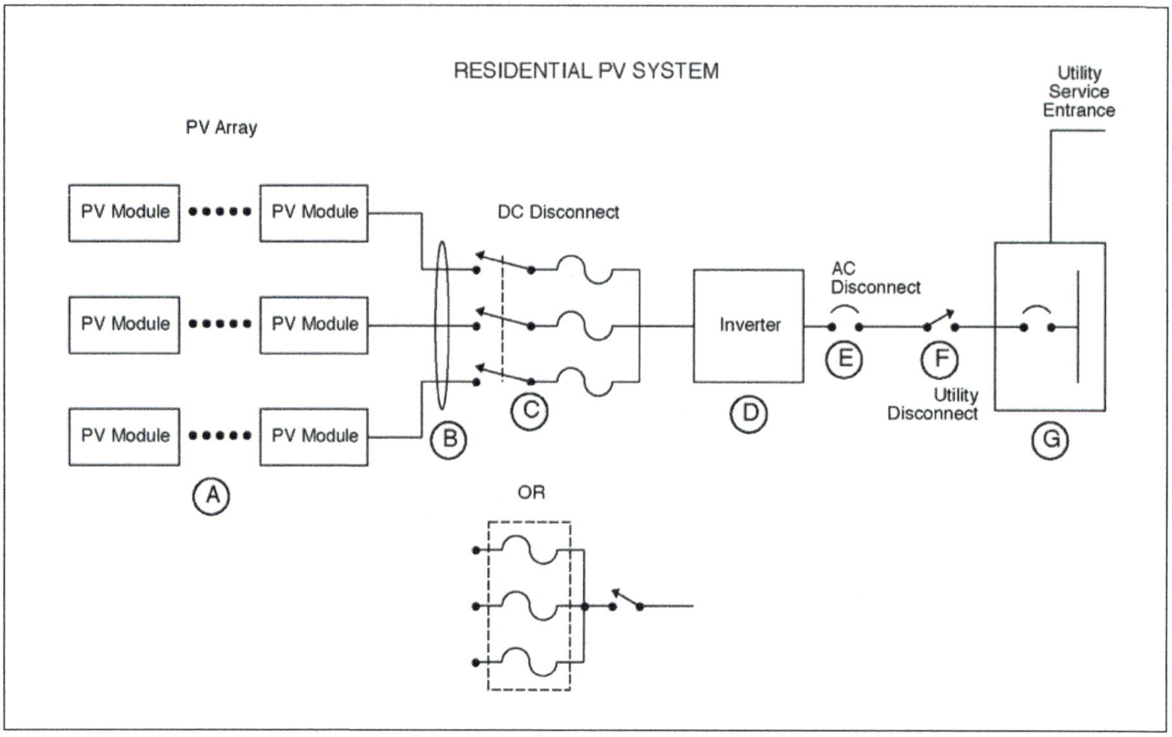

Figure 8.2 • One-line PV system diagram

pages from instruction manuals are the preferred way to present this information.

Since the PV modules and inverters are constantly changing, the preferred option would be to include copies of the installation manuals for the inverter, module, and dc PV combiner with each submission. Since not all module installation manuals include the technical specification, the technical specifications should also be included.

The Diagram

A one-line diagram such as shown in figure 8.2 should accompany the permit application. Since the details of disconnects and grounding are not familiar to all involved, a three-line diagram would be even better as shown in figure 8.3. While a formal CAD-generated diagram on 24" x 36" paper is not required, something better than a "back of the envelope" sketch should be presented. The circled letters in the figures will be referenced below to indicate information that should appear on or be attached to the plan.

This diagram should be accompanied by several additional pieces of information. An installation instruction manual for the PV module should be included along with a specification sheet for the module since frequently the specifications are not contained in the instruction manual. An installation instruction manual for the utility-interactive inverter should also be included along with an instruction manual for any dc combiner, where such a combiner is used to combine source circuits. Cut sheets should be provided to show the ratings of all disconnects and overcurrent protective devices used in the system. Information should be provided on the ratings of any existing electrical equipment such as load centers (main breaker and busbar rating) that the PV system will be connected to or that PV ac currents may pass through all the way from the PV inverter output to the service entrance conductors.

The System

On the one- and three-line diagrams, the following information should be indicated, or that

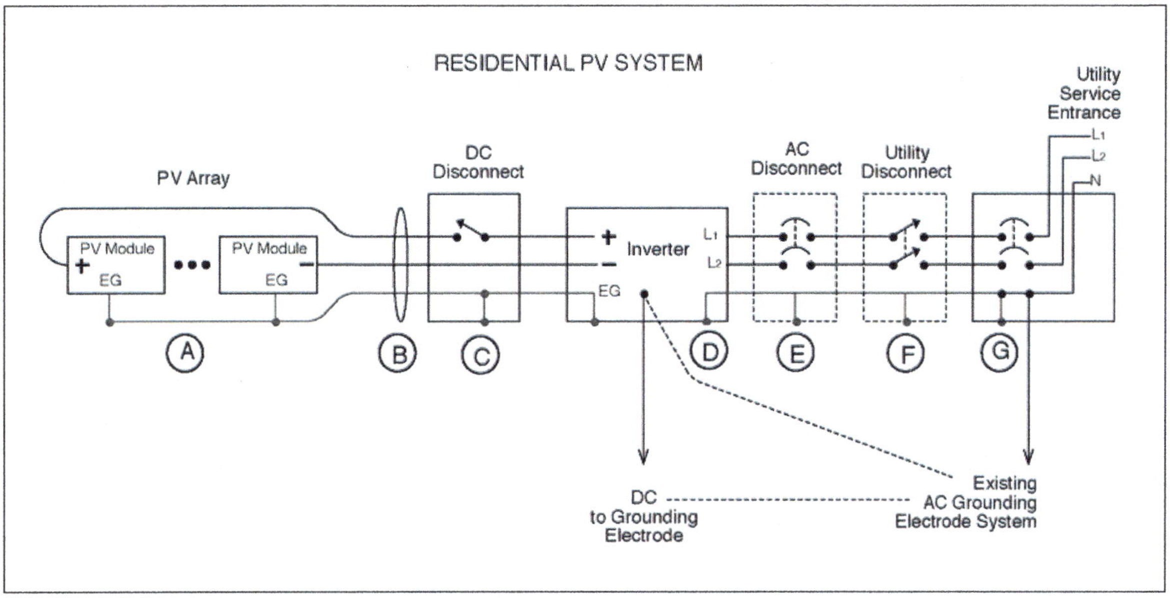

Figure 8.3 • Three-line PV system diagram. Either the ac disconnect or the utility disconnect can serve as the PV system disconnect.

information should be attached. The following numbers reference items in figures 8.2 and 8.3.

A. PV Array

A.1. The type and number of PV modules in each series string should be indicated. The open-circuit voltage (V_{oc}) of each module, times the number of modules connected in series, times a cold temperature factor (690.7) equals the maximum systems voltage and must be less than the maximum direct current (dc) input voltage of the inverter and less than the voltage rating of connected direct-current (dc) equipment (wires, overcurrent devices, disconnects).

The coldest expected ambient temperature should be noted on the diagram. A label on the back of each module as shown in photo 8.3 will give the electrical parameters needed for the code-required calculations. These values should appear on the diagram. A photo of the label should be included since this label will not be visible once the modules have been installed.

A.2. The ampacity of module interconnection cables, not less than the larger of: 1.56 times the module I_{sc} ***or*** conditions of use applied to 1.25 I_{sc}. Due to the exposed, outdoor location and high operating temperatures, all conductors should have insulation rated for 90°C and wet conditions (in conduit, THHN/THWN-2, XHHW-2 or RHW-2). Exposed conductors (USE-2 or PV Cable/PV Wire) must be suitable for the hot, wet, outdoor environment.

B. Conduits

B.1. Conduits will typically be used throughout the system and specifically after the dc wiring leaves the PV array and runs through the structure. They will be installed in various locations, some of which may be in sunlight. See 310.15(B)(2) and 310.15(B)(3) for temperature calculations and adders. The average high and the height of the conduit above the roof should be specified. Conduit fill and conductor ampacity calculations for conduit fill and temperature calculations should be included or attached.

One source of temperature correction factors for conduits in sunlight based on the 2014 *NEC* (still useful, if not specifically required in the 2017 *NEC*) is Copper Development's web site:

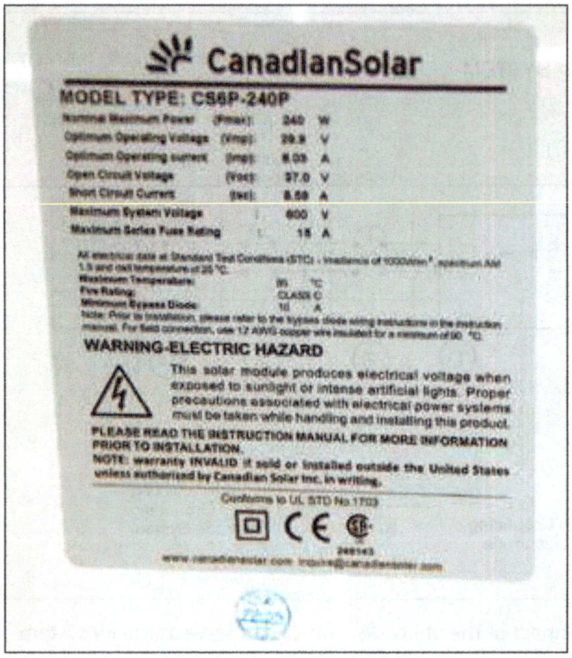

Photo 8.3 • PV module label showing electrical ratings

Photo 8.4 • DC PV disconnect with required markings. 2017 *NEC* requirements are slightly different. See 690.13 and 690.53.

http://www.copper.org/applications/electrical/building/pdf/rooftop.pdf

B. 2. The dc PV source circuit and dc PV output conductors must remain outside the structure until they reach the readily accessible PV dc disconnect unless the conductors are installed in a metallic raceway [690.13, 690.31(G)]. The exception for the use of metal conduits inside the building does not apply to the ac output of utility-interactive inverters since these circuits are similar to ac branch circuits.

C. Module and String Overcurrent Protection and PV DC Disconnect

C.1. Overcurrent protective devices (OCPD) in dc circuits may not be required when there are only one or two strings of modules connected in parallel to the inverter. Three or more strings of modules typically require an OCPD in each string. The current rating of the OCPD, when required, should be 1.56 Isc for that circuit (690.8, 690.9). The voltage rating of the OCPD should be not less than the maximum PV systems voltage found in A.1. The strings may be combined in parallel in a combiner box ahead of an unfused dc PV disconnect or combined at the output of the dc PV disconnect (figure 8.2 and photo 8.4). See chapter 3 for a more detailed look at the calculations and the requirements for OCPD in the dc PV array circuits. Any OCPD connected in series with a module or string of modules should not have a value greater than the maximum series fuse value marked on the back of the module (photo 8.3).

C. 2. The PV array output should be connected to the top or line side of the main dc PV disconnect. The circuit from the dc PV disconnect to the inverter dc input should be connected to the bottom or load side of the disconnect. Both ac PV output conductors must be switched by the disconnect except in those very small solidly grounded PV systems.

C. 3. PV output conductors, after any combining of series strings, should have an ampacity of not less than the larger of: 1.56 times the module I_{sc} times the number of strings in parallel ***or*** conditions of use applied to 1.25 I_{sc} times the number of strings in parallel. The overcurrent device must always protect the conductors. Additional details on sizing conductors and overcurrent devices will be found in chapter 6.

Module Grounding

Starting with the PV module, a grounding detail that is going to be used to ground the module

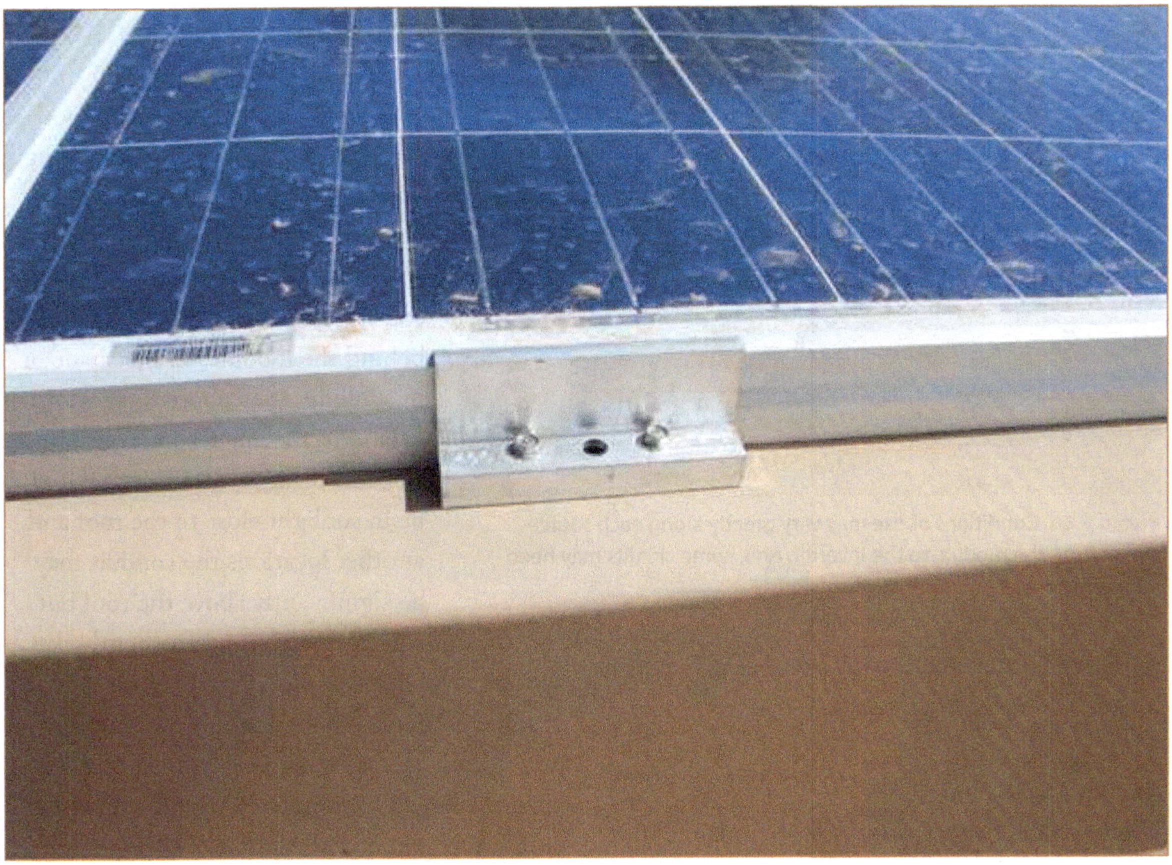

Photo 8.5 • Top clip mounting system must be mentioned and installation locations identified in the module instruction manual to keep the module listing valid. Note that drilling two extra holes in this clip has violated the listing on the clip.

frames should be provided on or with the diagram, and this grounding detail should be consistent with the grounding instructions provided in the module instruction manual. If the module *is not* grounded in a manner consistent with the module instruction manual, the listing on the module will be invalid. Since the *NEC* requires that the module be certified/listed [690.4(D)], then the installation of an unlisted module no longer meets *NEC* requirements. A frequent problem in this area is the use of some sort of a grounding device that does not use the marked grounding holes shown in the module instruction manual and the instruction manual makes no mention of grounding the module at positions other than the marked grounding holes.

Module Mounting

In a similar manner, a mounting detail for the PV module should be included with the diagram. This mounting detail should be consistent with the mounting instructions for the PV module. Again, if the module *is not* mounted according to the module instructions, the listing is invalid, and the module installation does not comply with the requirements of the *NEC*. The use of a mounting device or racking system that has top clips to mount the modules has become common recently (photo 8.5). The module instruction manual must show the use and location of top clips for mounting the module if this mounting system is to be acceptable and the installation to remain code-compliant. This alternate mounting system will be in addition to the normal mounting holes that are drilled in the back frame of the module that are normally used for mounting. Although not required by the *NEC*, the racking system may be listed to UL Standard 2703 that governs

Photo 8.6 • Conditions of use may vary greatly along each source circuit from the module to the inverter. And, some circuits may need additional protection from Mother Nature.

the mechanical and electrical grounding requirements for the racking system.

Racking Systems

Underwriters Laboratories (UL) Standard 2703 details a list of performance requirements for PV racking and mounting systems. Although the *Code* does not require racking or mounting systems to be certified/listed, a racking system that complies with this Standard will ensure that the racking system is structurally robust, and that the racking system can be used as an equipment grounding conductor.

The certification/listing of the racking system to UL 2703 *does not*, however, automatically guarantee that the racking system and its module mounting, and module grounding provisions meet code requirements. At this time, both the racking system instructions *and* the module instructions must indicate that these two devices have been evaluated together as a system in order to ensure that the mounting of the module to the racking system and the grounding of the module to the racking system comply with all provisions of both standards and, therefore, will meet the requirements of the *NEC*.

Conductor Conditions of Use

The conductors from the PV modules to the inverter will pass through several different areas on or in the building; each area will have a different set of conditions of use. At some places, they may be in free air behind the modules in the shade or in sunlight. (See photo 8.6.) At other locations, they may be in conduit. In some locations, the conduit will be in sunlight close to the roof and at other locations the conduit may be significantly above the roof but still in sunlight and frequently the conduits will be in shade on the roof. Usually, at some point, the conduit will be inside conditioned space as it is routed to the inverter. The diagram should be annotated to show these different conditions of use and the temperatures and spacing from the roof expected in each segment of the conductor run. Of course, the worst-case conditions of use will be applied to determine the conductor size. It should be noted that, in some installations, the conductor size for the field-installed conductors may exceed the size of the conductors attached to the modules themselves because the conditions of use for the field-installed conductors and conduit in sunlight may be more severe than those associated with the conductors attached to the modules which are typically in free air. An ampacity calculation showing the conditions of use and the required conductor size should be shown for each separate condition of use or at least the worst-case conductor size and that calculation should be shown (310.15).

D. The Inverter

D.1. The inverter must be listed for utility-interactive (U-I); use 705.6.

Photo 8.7 • Inverter with internal ac and dc disconnects. These may meet the requirements for inverter isolating devices (690.15)

D.2. The inverter maximum input voltage must not be exceeded in cold weather [110.3(B)].

D.3. Most PV systems must have a dc ground-fault protection device (GFPD) [690.41(B)]. When a GFPD is built in to the inverter (nearly all U-I inverters), there should be no external (to the inverter) bond between the grounded circuit conductor and the grounding system. Additionally, an equipment grounding conductor must be routed from the PV array through the dc PV disconnect all the way to the inverter.

D.4. In addition to ac and dc equipment grounding conductors, the inverter providing a PV system functional ground may have a requirement for connecting the ac equipment-grounding conductor [690.47(A)]. The dotted lines in figure 8-2 show alternate routing and bonding for the dc grounding electrode conductors (required by the 2014 *NEC*).

Photo 8.8 • Inverters installed with external ac and dc disconnects

Photo 8.9 • Utility-required ac disconnect. Inverters are located at the ground-mounted PV array.

D.5. DC and/or ac disconnects internal to the inverter are acceptable if the inverter is readily accessible and the AHJ judges that only qualified people will service the inverter (photo 8.7). Internal dc disconnects are common in the smaller inverters. Otherwise, external disconnects will be needed (photo 8.8). Internal disconnects, if circuit breakers, will not protect the ac output circuits from utility-sourced fault currents and an external OCPD will be needed at the utility point of connection.

E. Inverter AC Output Overcurrent Device and Disconnect

E.1. Any OCPD located in the inverter ac output should be rated at 1.25 times the rated continuous output current of the inverter. The rated continuous current is specified in the inverter manual or is calculated by dividing the inverter rated output power by the nominal ac line voltage. This OCPD may be a backfed breaker located in the dwelling load center, the place where any possible fault currents for the inverter ac output conductor would originate. A backfed breaker in the dwelling load center could also be the PV system disconnect and the inverter ac disconnect (690.13, 690.15) if the inverter were located near the load center.

E.2. The inverter ac disconnect should be "grouped" with the dc inverter disconnect and both should be "near" the inverter. The AHJ determines the definitions of "grouped" and "near." Most systems use the PV disconnect as the dc inverter disconnect, but if the PV dc disconnect is on the outside of the building (no longer a requirement in the 2017 *NEC*) and the inverter is on the inside, a second dc inverter disconnect may be required inside the building at the inverter location. The same thing would apply if the backfed circuit breaker in the building load center were on the outside wall and the inverter were on the inside. A disconnect (usually a circuit breaker) would be required inside the building near the inverter.

E.3. From the above, it becomes obvious that the system diagram should show the approximate physical location of all components.

At the DC PV Disconnect and The Inverter

The inverter instruction manual should be consulted to determine any instructions for implementing the PV system functional ground.

Inverter AC Output Circuit

The conductor from the inverter output to the first overcurrent device must meet the requirements of *NEC* Section 690.8 for size. And, the overcurrent device itself should be rated according to the requirements of Sections 690.8 and 690.9 of the *NEC*. The conductor sizes for the ac output circuit of the inverter including any conductor downstream of the inverter *after* the first overcurrent device that may carry both inverter output current and utility current must comply with *NEC* Section 705.12(B). Someplace in the permitting package, the calculation should be shown for each circuit involved all the way back to the service-entrance equipment. Backfed circuit breakers in each panelboard or load center that will handle PV current should be marked on the diagram showing their correct location, which in most cases will be at the opposite end of the busbar from the main breaker or main lugs for that panelboard. Chapter 7 provides additional details on the various PV system-to-utility connections.

F. Utility-Required AC Disconnect

F.1. Many utilities require a visible-blade, lockable-open disconnect in the ac output circuit of the inverter. This disconnect is usually located within sight of the service-entrance meter so that emergency response people can easily find it. The top terminals (line side) of this disconnect should be connected to the circuit that comes from the utility (supply-side connections, [705.12(A)]) or the ac load center (load-side connection, [705.12(B)]) because they will usually be energized by utility voltage. The bottom terminals (load side) should be connected to the circuit from the inverter. This disconnect may be fused or unfused depending on the specific requirements of the utility and its location in the circuit. Photo 8.9 shows an ac disconnect between the REC meter and the revenue meter that serves as both PV system disconnect and the utility-required ac PV system disconnect. The utility point of connection is inside the house through a backfed circuit breaker in the load center.

F.2. The utility disconnect must have a minimum current rating of 1.25 times the maximum continuous output current of the inverter (690.8) and be protected by the downstream (toward the utility) OCPD. For a supply side connected PV system [705.12(A)], the disconnect must also be rated as suitable as service entrance equipment [690.13(C)].

Point of Connection-Load Center

G.1. Most of the smaller residential and commercial PV systems will make the point of connection with the utility through a backfed breaker in the building. *NEC* Section 705.12(B) establishes the requirements. See chapter 7 for the details of these connections.

The Inspection

With the some of the *Code* requirements verified during plan review, the field inspection will usually go much smoother. *Code* requirements that are found not to be met on paper may be flagged for the installer, and they should be corrected before taking valuable time going to the site and attempting to ensure that everything meets *Code* by a visual inspection of the installation. I have reviewed several hundred plans for PV systems over the last 30 years (5 kW to 4 MW) and note that many systems fail to meet all *NEC* requirements on paper. The systems that *do* comply with the *NEC* requirements are usually plans from PV installers/systems integrators that have been in business a number of years and have been installing similar systems over that period.

Summary

All the above information should be included in plans submitted for obtaining a permit for the installation of a PV system. The more information submitted, the easier it will be for the PV system designer/installer to communicate to the inspector/permitting official that the system design and component selection meet the requirements of the *NEC*. It is far more cost effective to change the design on paper before any hardware is purchased and installed than it would be after the system has been installed. Ready for the inspection? Read on.

09

The Process of Inspecting PV Systems

While large, utility-scale photovoltaic (PV) power systems account for more of the installed megawatts and gigawatts of PV power in the United States, the typical inspector will be inspecting far more residential and small commercial PV systems than utility-scale PV systems. While failures in those utility-scale PV systems can be spectacular, these systems are usually ground-mounted in remote areas, and damage is restricted to the equipment itself, and of course the pocketbooks of organizations involved directly with the system.

On the other hand, when residential and small commercial PV systems fail, the resulting damage can affect homes, commercial buildings, and people (including first responders) in a very real manner. To ensure the safety of the public, the inspection community including administrative personnel and managers, as well as the people who are responsible for funding must work together to provide an exceptional final PV inspection before the system is turned on. That final inspection will ensure the safety of an electrical power system that may be generating energy for the next 40 or 50 years, possibly without any further attention.

The Way It Should Be Done—*The Author's Opinion*

The Overall Process. The inspection process for a PV system should begin with the permitting process that should include a full-system description which shows a three-line diagram and manuals for all PV-unique equipment being installed. Chapter 8 addresses the details. Hopefully, permits will be issued only to individuals or organizations who are qualified to install PV systems and have the necessary licenses. In the best of circumstances, the permit applicant will be well-

Photo 9.1 • Some PV systems may experience trouble, now or in the future, when installed by less than competent organizations.

Photo 9.2 • Hands-on field inspection training for AHJs

versed in installing residential and commercial electrical systems, with additional training in PV code requirements and installations.

A plan review step in the comfortable confines of an office environment should be performed on the permit application material. Unfortunately, many aspects of the PV installation cannot be verified after the system is installed. Modules will be mounted close to a roof, and the labels cannot be seen. Conductors may be installed in not readily accessible areas, and sizing and types will be difficult to verify.

The initial inspection should verify that all requirements in the *NEC* have been met and that the equipment that has been installed agrees with the information provided in the permit application. Typically, there is no utility interface connection at this time, and several parts of the system operation cannot be evaluated for safety.

In most cases, the initial PV inspection will be followed by a utility inspection where ac electrical power is connected to the system by the utility. After this is done, a subsequent inspection should be made to verify that equipment like PV Rapid Shutdown systems and the like are functioning as intended.

First Comes Training and Education

It is imperative that the inspector of PV systems to get adequate knowledge and training so that he or she can effectively inspect PV systems to ensure the safety of the public. Acquiring this background is not as easy as it sounds (photo 9.2).

For example, the author is involved in conference calls a couple of times a week dealing with the standards associated with PV systems. New codes and standards are reviewed on a regular basis and assistance is provided in writing the same. Even the author has trouble keeping up with the changing technology, the changing standards, and the changing codes associated with PV.

As inspectors, we must take it upon ourselves to get as much training in this ever-changing technology as possible. There are online training programs, and there are hardcopy materials that can be used. The first example that comes to mind of hardcopy material (or electronic) is the *National Electrical Code Handbook* (*NECH*). In many cases, the informative material found in the *NECH* is based upon the background and technical information that went into developing the code requirement. Every inspector should have a copy of the Handbook and review this document almost continually. It takes nearly three years to gain familiarity with the current version of the *Code* and then the new edition is released.

Another source is these "Perspectives on PV" articles written by the author that are archived on the IAEI website. We should also review the IAEI Publications Catalog (http://www.iaei.org/web/store) for all the significant documents that summarize the new code requirements. These documents are updated every code cycle, and the IAEI Publications Catalog is probably the best source for a broad spectrum of documents that can be used in educating the inspector on the nuances and changes in the *Code*.

Another source of wide-ranging technical articles on PV systems, PV equipment, and PV installations can be found on the Solar America Board of Codes and Standards (solarabcs) website (http://www.solarabcs.org/).

One source of online training that is widely advertised in *IAEI Magazine* and other publications is the "PV Online Training Course for Code Officials" developed by the Interstate Renewable Energy Council (IREC), IAEI and others. It can be accessed here: www.pvonlinetraining.org. Be advised; a game is included in this training program. See the excellent article in *IAEI Magazine* (Nov-Dec 2016) on this program (https://iaeimagazine.org/magazine/2016/11/04/become-solar-smarter-with-pv-training-online/).

Building Officials and Administrators. Inspectors must be provided with adequate funding and time allowances for the necessary study and training. Workshops and presentations may be brought to a central location so that many inspectors can participate, or video conferencing/webinars can be set up to ensure that all inspectors maintain currency in their respective disciplines, in this case, PV systems.

The Permit

The permit for a PV system should include more than a simple form and the payment of a fee. That permit application should contain the following:

a. A three-line diagram of the system showing, as a minimum, conductor sizes, conduit sizes, equipment grounding circuits, disconnects, overcurrent protection, and the location and method of making the utility interconnection.

b. The code calculations for the maximum system voltage including the expected lowest temperature at the site, the ampacity calculations for each circuit including the ambient temperatures used, and the *NEC* 705.12 calculations for the interconnection point.

c. A copy of the installation manuals for the PV module, the dc combiner, the inverter, the PV rapid shutdown system/equipment and any other PV unique equipment, as well as, cut sheets for the various disconnects and any load centers.

d. PV module mounting and grounding details showing that these installation details are compatible and comply with the instructions in the PV module and the mounting rack manuals.

Requiring this material in the permit application will help to ensure that qualified persons are making that application. The material also will be useful in the plan review process (described below) that will facilitate the actual on-site inspection.

Expedited Permitting. For those jurisdictions not having a formal permitting process, an expedited permitting process developed by the Solar American Board of Codes and Standards (SOLARABCS) and written by Bill Brooks and others can be found here: http://www.solarabcs.org/about/publications/reports/expedited-permit/

This expedited permitting process is very useful for PV systems of 10 kW and less and includes diagrams, fill-in forms, and code-based calculations. It has been used by numerous jurisdictions throughout the country and modified for use with additional systems by other jurisdictions.

The Plan Review

Advantages. One advantage of having the details described above submitted with the permit application is that a quick review of this material will give some indication of the capabilities and competencies of the organization making the application and, hopefully, the person installing the system. If the basic system design in the permit package does not meet code requirements, it is almost certain that the system will not be installed in a safe and code-compliant manner. Another advantage of performing a plan review is the fact that the review can verify partial code-compliance in the warm, quiet, office environment and not out in the cold, blustery or hot, sunbaked environment of a rooftop PV system.

As mentioned above, it is physically not possible to fully access all the labels and markings on the PV modules and many of the cables after the PV array has been installed. The PV arrays are, in many cases, mounted within a few inches of the roof and it would require unmounting a PV module to view the data label on the back. Depending on the module grounding method being used, grounding hardware might have to be replaced when a module is removed from the racking system. Conductors installed under the modules and the connectors used between the PV modules and the field-installed wiring may not be easily accessible.

Additionally, reviewing the system on paper allows the AHJ to see the overall system and how the various components are interconnected. It will also show the flow of power through the system which will assist in determining if disconnects and overcurrent protection devices have been located properly.

To a certain extent, some of this information is provided by a full permit package and can be used to verify code compliance. Of course, this assumes that the equipment and materials in the permit package match the equipment and materials that will be installed; and that may not always be the case. This item should appear on the field inspection checklist.

Items to Look For. Here is a brief checklist of items that should be verified during a plan check. Significant numbers of items that are not code-compliant may signify that the basic PV system design is inadequate and that the permit package be returned to the submitter for corrections before any field inspection can be undertaken. A more detailed checklist will be found in Appendix A.

- All equipment listed for the specific application [690.4(B)]
- Modules and racks compatible for mounting and grounding [110.3(B)]
- Calculated maximum system voltage less than rated maximum equipment voltage [690.7]
- Cable types suitable for the environments, suitable for the circuit, and properly rated [690.8]
- Conduit and raceway selections appropriate for the environments and code-compliant [Chapter 3]
- Conduit fill calculations correct [Annex C]
- Disconnects suitable for the type system-grounded or ungrounded
- Disconnects properly rated and properly located [690 Part III]
- Overcurrent devices properly rated and properly located [690.9]
- Utility point of connection properly rated and located [705.12]
- Any utility-required disconnects and meter properly rated and located
- Equipment grounding and system grounding circuits correct
- PV Rapid Shutdown System equipment located in circuits where required by instructions [690.12], [110.3(B)]
- DC PV Arc Fault system incorporated into the PV system [690.11]
- DC combiners properly rated
- Inverter and inverter circuits properly rated for dc inputs and ac output

The Onsite Inspection

First impressions are important. Have you worked with this PV installer or systems integrator before? Is the site uncluttered and the system ready for inspection? Did they make the roof accessible in a safe manner? From a distance, does it look like good workmanship has been employed?

As with any electrical power system, attention to details during the installation process is critical, and the overall first impressions give some indication as to whether the installer has exercised good work procedures and habits.

Equipment installed to match the permit? The equipment that is installed should match the equipment that was listed in the permit application package. Variations in any components such as the PV modules or the inverter may necessitate recalculation of various parameters and the replacement of conductors, overcurrent devices, and other equipment.

Let the installer operate the system. From a liability and safety point of view, it is usually a good idea to let the installer of the system perform all handling of the system components and operation of those components during the inspection (photo 9.3).

The installer should have the necessary tools and test equipment available to open equipment, make measurements, and exercise the equipment in various modes to demonstrate compliance with *Code* requirements. Unfortunately, the PV systems have reached a level of complexity that it is no longer possible to verify full code-compliance by just looking at a static assembly of wires, conduits, circuit breakers, PV modules, inverters and other equipment. In most cases, the equipment must be operated to demonstrate the proper safety features and compliance with *NEC* requirements. In many cases, the utility will require an inspection by the local jurisdiction before they make the interconnection and they may necessitate a second inspection by the AHJ of the operational system.

The installer should be able to show the torque screwdrivers and torque wrenches that were used in making the electrical connections and to show where the torque values for the various pieces of equipment can be found. All electrical contacts that are bolted or screwed must be torqued to the proper value to ensure a safe and durable connection. [See 110.3(B) and photos 9.4 and 9.5.]

Proper grounding is critical for the long-term safety of the system. Before touching any parts of a new PV system, the equipment grounding systems from the PV modules to the main service panel and the grounding electrode(s) should be verified by visual inspection. The installer might be asked to show that various bolted grounding connections have been properly torqued, and that conduit nuts and grounding bushings are tight. Grounding/bonding bushings should be used on all metal conduit fittings operating over 250 volts ac or dc (250.97). (See photo 9.6.)

Photo 9.3 • Inspecting PV systems can result in unpleasant surprises.

On to the roof—Or not. What are some AHJ options when the local jurisdiction does not permit personnel to climb on the roof? The installer should be asked to provide close-up pictures of the PV module mounting and grounding methods and devices used. The pictures should show that the instructions for the module and rack mounting system were followed.

Although not required by *Code*, some jurisdictions are requiring the use of UL 2703 certified rack/mounting systems on the roof of residences. Since these racking systems cannot directly guarantee the quality of the actual attachment to the structure of the house, the author feels that any well-assembled module mounting and grounding system should suffice.

The same considerations might be given to ground-mounted installations where Unistrut embedded in concrete-filled holes might be judged an adequate mounting system. For ground-mounted

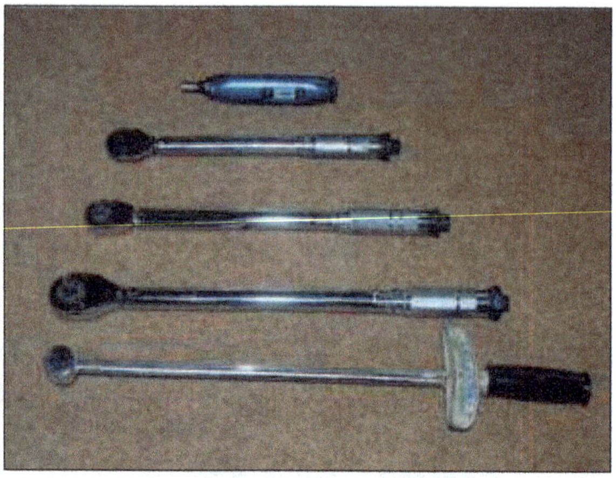

Photo 9.4 • Various sized torque tools are necessary for making code-compliant electrical connections.

Photo 9.5 • Improperly torqued splicing devices have melted.

Photo 9.6 • Metallic conduit operating at 400 volts dc connected to a concentric knockout with no grounding/bonding bushing. Grounding lug (not listed for this enclosure) attached with sheet metal screw violates Code and disconnect listing.

systems that are readily accessible, the exposed single-conductor module wiring must be made "not readily accessible" by fencing or guarding [690.31(A) and photo 9.7].

Instructions supplied with new modules will have the following phrase or equivalent: "This module remains certified to UL Standard 1703 only when mounted and grounded according to the included instructions."

A picture of the back or underside of the array will indicate if the loose, single-conductor module wiring has been properly secured. Any conductors not well-secured for decades of use will be subject to wind-driven movement and possible abrasion of the insulation, creating a potential shock or ground-fault hazard in the future (photo 9.8).

Another relatively recent issue in PV installations is the fact that the mating pairs of connectors used on modules and field-installed USE-2 and PV wire must be from the same manufacturer and of the same series. Even though connectors from two different manufacturers are said to be "MC-4 compatible," they have not been tested and certified/listed with each other and therefore it would be a violation of the connector and module listing to use them together (photo 9.9). See chapter 3 for additional details.

Back on the Ground

At the utility point of connection. In a 705.12(B) utility-interactive inverter output connection, the rating and location (position on the load center busbar) of the backfed PV breaker should be verified. Note that the 2017 *NEC* uses the term *interactive power source* or just *power source* when referring to the utility-interactive inverter. The 120% or 100% rule should be met, and for the 120% rule, the backfed PV breaker(s) should be at the opposite end of the busbar from the main breaker (photo 9.8). Other sections of 705.12(B)(2) may be used and these should be substantiated with the appropriate calculations.

Many residences and commercial buildings have subpanels. Where subpanels are involved, attention must be directed to ensuring that any feeders or main-lug-only subpanels have appropriate ampacity or ratings to deal with backfed PV currents in the main panels adding to the currents from the main service breaker. In some cases, the feeder may be protected by a breaker at its source end, and the subpanel may be protected by a main breaker in that subpanel. If these protections are not afforded, the ampacity of the feeder and the rating of the subpanel will have to be increased. Having the backfed PV breaker mounted in the last subpanel and in the last position in that subpanel will avoid some complications, but feeders and the main panel must be checked for *Code* compliance in Article 705.

For a supply 705.12(A) supply-side connection, the rating of the overcurrent protective device should be checked, and the location and wiring (utility connection to the top of any safety switch) of any utility required or used ac PV disconnect verified. The overcurrent device connected to the output of each inverter may not exceed the maximum overcurrent protective device as specified in the inverter installation manual; yet another reason to have the manual available during the inspection. The presence of the required overcurrent device within 10 feet of the service entrance connection point (705.31) should be verified.

It is the opinion of the author, since those (unnamed) conductors between the actual connection to the service-entrance cables and the first overcurrent protection device are subject to the full available fault currents from the utility and because those conductors may be smaller than the service-entrance conductors, they should be treated as service-entrance conductors per the requirements of Article 230 with respect to routing and mechanical protection.

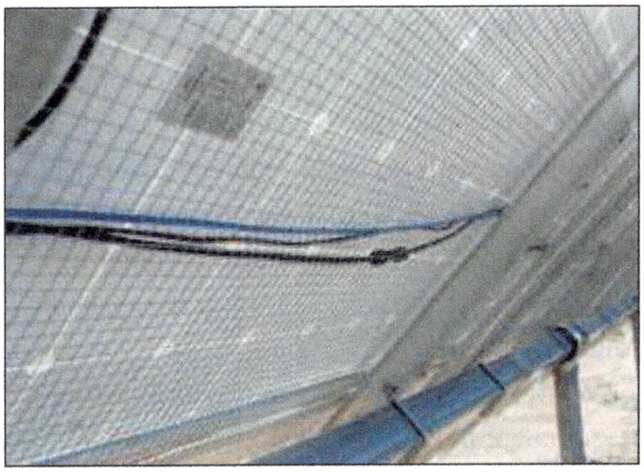

Photo 9.7 • Exposed, readily-accessible module wiring "guarded" by grounded, galvanized hardware cloth attached to back of module racks.

Photo 9.8 • Improperly secured module wiring will be subject to future wind-driven damage.

Photo 9.9 • Bottom: Connectors from the same manufacturer/series properly paired for a code-compliant installation. Top: Connectors from two different manufacturers improperly paired, resulting in a certification/listing violation, a Code violation and a potential safety hazard.

After the System is Powered

With the complexity of modern PV equipment, full verification of the installation for safety and compliance with the *NEC* requirements (including Section 110.3(B) that requires compliance with equipment installation instructions) will usually require the AHJ to inspect the system during operation.

Verify equipment adjustments and settings. Equipment like string inverters, multimode inverters, charge controllers and PV Rapid Shutdown Equipment (PVRSE) may have software/firmware adjustments that must be properly made to ensure the safe and correct operation of the product [110.3(B)]. The AHJ should have the installer demonstrate that all adjustments have been completed in accordance with the product instruction manuals.

PV Rapid Shutdown System (PVRSS).

The inspector should have the installer demonstrate the proper operation of the PVRSS since this system is directly involved with life safety issues for first responders. The ac power to the building that the PV is installed on should be turned off, as it would be during any first responder activity. In fact, the main service disconnect or the ac PV disconnect may have to be marked as a second PVRSS initiator in some systems to ensure the ac inverter output circuits are controlled by the PVRSS in addition to the dc circuits. The main ac service disconnect for the building may or may not be the PVRSS initiator. In any event, the ac output circuits of the PV system should not remain energized after the PVRSS system has been initiated and the facility main service disconnect has been opened.

AC PV module or microinverter system. In the case of an ac PV module system or a microinverter system, the installer should connect a digital voltmeter (DVM) to the ac output of the operating ac PV module array with line voltage indicated. The installer should then activate the PVRSS with the well-marked initiator, and the AJH should verify that the measured voltage drops to the required value of 30 volts within 30 seconds. Note: at some future date, the ac voltage level may be reduced to 15 volts due to proposed changes in UL Standard 1741. Both line-to-line and line-to-ground measurements will be required.

String inverter systems. The string inverter system will usually require that the dc conductors from the array and the ac conductors from the inverter be controlled by the PVRSS. Some inverters may be listed as PV Rapid Shutdown Equipment (PVRSE) and will meet the voltage requirements on dc inputs and/or ac outputs without external equipment. Appropriate meters should be connected to the dc input conductors that go to the inverter (normally, at the closed dc disconnect) and to the ac output conductors of the inverter. The installer should activate the PVRSS initiator, and the AHJ should verify that all voltages go to the required 30-volt levels within 30 seconds. AC voltage requirements may be reduced to 15 volts due to future changes in the UL standard. Line-to-line and line-to-ground measurements will be required.

Arcs Will Happen

As the PV system ages, it is expected that there will be greater opportunities for series arcs to form in the dc array wiring and the modules. Connectors, even those properly mated in pairs, may loosen over time from wind driven vibrations where conductors have not been securely fastened. There will be some connector pairs that are mated from different manufacturers that get past the inspections and may create failures sooner rather than later. Module solder bonds have failed in the past, and with numerous new manufacturers of PV modules on the market, we may expect to see such failures in the future.

The requirements for PV system to have dc PV arc-fault circuit interrupter (DCPVAFCI) are similar, but not identical in 2011, 2014, and 2017 *Codes*. Verification that the inverter, dc PV combiner or charge controller has the DCPVAFCI will be accomplished by the markings on the inverter, dc combiner or possibly the charge controller. At this point, it is unclear how the UL Standard 1699B will deal with products that must meet the differing requirements of various *Code* editions. It is possible that an installer controlled software function will allow a single inverter to meet the varying requirements of several editions of the *NEC*.

Ride-Through PV Inverters

Various states and utilities are starting to require the use of utility-interactive inverters that will stay on line and produce full available array power during utility voltage and frequency variations that exceed the normal anti-islanding limits of +10% to -12% on voltage and +0.5 Hz to -0.7 Hz on frequency. The voltage variations may be as low as 50% of the normal line voltage. These inverters will also be required to come back on line at full array power within one (1) second after the utility voltage and frequency have returned to either the normal or widened limits. This is considerably different from the current five (5) minute delay in powering up after a utility outage. There will be certified/listed utility interactive inverters on the market that will meet these new requirements and they will be well marked as such. However, the great majority of the available inverters will have the current anti islanding software/firmware. It will be up to the AHJ to determine that the correct inverter has been installed based on the local requirements. It will not be possible on the typical PV inverter to easily test these characteristics in the field, so the markings and the certification/listing mark will have to suffice.

Summary

Photovoltaic power systems are proliferating and are going to be a major player in our nation's energy mix. They are going to become more complex. The inspector community must be able and willing to inspect these systems as the last line of defense in ensuring the safety of the public. The plan review stage is becomisng increasingly important due to the increasing complexity of the systems and the number of not-fully-qualified individuals and organizations installing these systems.

Those of us in the inspection community and those associated with the inspection community including inspectors at all levels, chief inspectors, building officials, and administrators responsible for funding need to work diligently to increase the competency and quality of our inspection process and our inspection force. Additional time must be made available to inspect these systems. Additional time and funding must be made available to educate and train the inspector workforce, and above all, those inspectors must be fully supported when they make the tough calls on code violations.

10

The 15-Minute PV Inspection–Can You? Should You

In some jurisdictions, inspectors have as little as 15 minutes to make a residential electrical inspection. A common question is, "Can I inspect a residential PV system in 15 minutes?" This chapter will examine that question and take up the question, "Should only 15 minutes be allocated for inspecting a residential PV system?"

Let's start with an ideal situation. The inspector is familiar with PV systems in general and has inspected quite a few. He or she receives an application for a permit for a PV system, and that application is accompanied by all the material outlined in the preceding chapters in this book. A plan review of the supplied material shows no major problems in code-compliance, and the installer quickly rectifies the few minor problem areas found. A team consisting of a PV vendor/installer with a history of good PV installations employing or working with an electrical contractor/electrician who has a commercial electrical license and some PV experience has done the design of the system and the installation. The system is nearly identical to others that this organization has previously installed and have been inspected with no significant issues. The system configuration and the equipment (modules, inverters, disconnects and overcurrent devices) that has been installed is the same as previous systems and those previous systems have been inspected and no significant issues were found.

Here are some of the items that an inspector should verify during the site visit. They are listed in order of importance and in order of safety for the inspector. For a more complete checklist, see the Inspector/Installer Checklist found in Appendix A.

Grounding

Proper grounding of the PV system is extremely important because the PV modules will be generating hazardous amounts of energy for the next fifty years or more. Proper grounding is the first, the last, and the most important area (in the author's experience) that requires code-compliance in a PV system. Proper grounding of all exposed metal surfaces that may become energized as the system ages or as accidents happen will provide the highest levels of protection against shock and fires. Proper grounding will also facilitate the action of the ground-fault protection and arc-fault protection equipment that these systems will have. As the inspector moves through the

Photo 10.1 • Improperly wired 240-volt inverter; neutral wired to PE terminal

As PV systems mature and UL standards and the *Code* evolve, it is hoped that the grounding of PV systems will become more robust.

AC Point of Connection to the Utility

While the premises ac load center is open to check the grounding connection, the location and value of the backfed PV circuit breaker can be noted. It should match the value on the permit application and shall not be greater than 20% of the load center rating. It should be located at the opposite end of the busbar from the utility input. This assumes that the main breaker and the load center have the same rating. See *NEC* 705.12(B)(2). This requirement limits the backfed PV breaker to a maximum of 20-amps on a 100-amp load center and to a maximum of 40 amps on a 200-amp panel. Breakers larger than this indicate that the utility connection should have been made on the supply side of the service disconnect. See Chapter 7 for more details.

PV system, grounding will be a critical inspection item in several locations.

Most smaller PV systems (below 10–20 kW) may have all the PV equipment, both ac and dc, grounded by a single "grounding" conductor connected from the modules to the grounding bus bar in the existing ac load center. The module frames, the PV array mounting rack, and the dc disconnect are connected with a dc equipment grounding conductor that connects to the inverter. From the inverter through one or more ac disconnects, an ac equipment grounding conductor (which may also serve as any PV system functional ground) continues the connection to the ground bus bar in the ac load center [690.47(A)].

The first item an inspector should verify is that the ac equipment grounding conductor from the PV system inverter has been connected properly in the ac load center grounding bus bar and that the ac load center has a proper connection to ground (earthed). If this equipment grounding *has not* been done properly, a ground fault in the PV array or elsewhere in the system may put several hundred volts (with respect to the ground where the AHJ is standing) on the ungrounded exposed metal surfaces of any PV equipment.

Inverters

The inverter should be opened to check the field-installed connections. Some inverters will require metric hex socket drivers (or Allen wrenches) to open. One manufacturer makes a sealed inverter with permanently attached cables for connections to the adjacent ac and dc disconnects.

Inverters with a 120-volt output should have line (ungrounded), neutral (grounded), and equipment grounding conductors between the load center and the inverter. Inverters made outside the U.S. may have the equipment grounding terminals marked PE for "protective earth." Some 240-volt inverters have only line 1, line 2, and equipment grounding conductors with no neutral (grounded)

Photo 10.2 • Inverter with two dc inputs (lower left). EGC on lower right. Improper double lugging for small surge suppressor conductors. Color coding based on 2014 *NEC*.

Photo 10.3 • Improperly grounded dc disconnect; violates 250.8, 110.3(3), 250.96(A), and 250.4(A)(5)

conductor, while others will have line 1, line 2, neutral, and equipment grounding conductors. The inverter manual (submitted with the permit request) will show the proper connections. Inverters requiring no neutral connection must not have the neutral terminal in the utility circuit attached to anything, particularly an equipment grounding terminal, because such a connection would make a double bond between ground and the neutral conductor that is prohibited by *NEC* Section 250.6 (photo 10.1).

The dc input connections to the inverter may include one or more sets of positive and negative conductors as well as at least one dc equipment grounding conductor routed to either an external dc disconnect or to the PV array (photo 10.2).

AC and DC Disconnects

Each disconnect should be properly grounded with the equipment grounding conductors or metal raceways. Following and verifying the equipment grounding conductors backwards from the ac load center through the system to the PV modules is important to ensure that each exposed metal surface that may be energized is grounded. Grounding using sheet metal screws is prohibited by the *Code* and the use of thread cutting screws and aluminum lugs is questionable (photo 10.3). Most listed fused disconnects and circuit breaker enclosures have ground-bar kits with specific mounting instructions and locations that should be used to maintain the listings of the devices and to provide the highest quality grounding connection (photo 10.4). Metal conduits with dc circuits operating over 250 volts will usually require grounding/bonding bushings (250.97).

While any of the disconnect or isolation device enclosures are opened, the color coding of the conductors should be checked. PV systems installed under the 2017 *NEC* will typically not have any of the dc PV array conductors grounded and no white conductors should be seen. The exception would be the small solidly grounded PV array outside a building as allowed by 690.41.

There is no specified color code for the ungrounded conductors, and any color is permitted as long as gray, white, green, and green and yellow are not used. In conduit, conductors with colored insulation can be used for polarity and circuit

Photo 10.4 • Properly installed, listed ground-bar kit

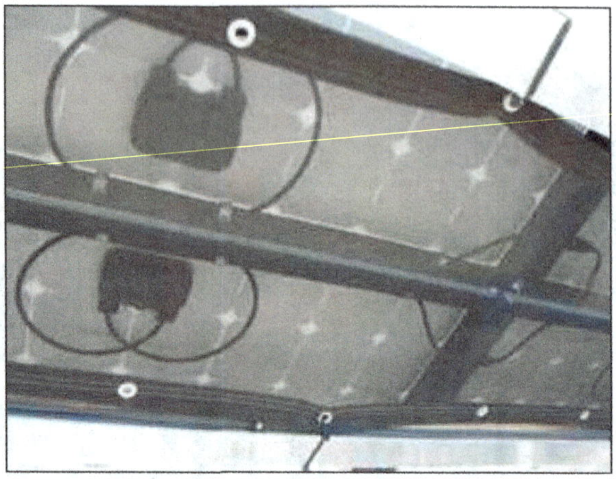

Photo 10.5 • Readily accessible module wiring properly secured and guarded.

identification. However, colored insulations on exposed wiring may not prove durable over the life of the system due to a lack of carbon black in the insulation resulting in a reduction in the ultraviolet (UV) radiation resistance. Black insulation has proven to be the most durable.

Both circuit conductors (positive and negative) should be routed through the disconnect enclosure even when only the ungrounded conductor is switched. Avoiding a "switch loop" configuration ensures that both circuit conductors are always in close proximity (lowering circuit inductance) for best functioning of overcurrent devices and to allow a bolted connection point for the grounded conductor on an isolated "neutral bus" in the enclosure, if required.

In the "switch-type" dc PV disconnect, the always "hot" conductors from the PV array wiring source or output circuits should be connected to the top (protected) "Line," terminals on the switch while the lower, exposed, "Load" terminals should be connected to the inverter. On any ac switched disconnect, the upper "Line" terminals should be connected to the utility power conductors that come from the backfed ac load center or a supply-side utility connection. The lower "Load" terminals should be connected to the inverter.

Workmanship and the Roof

The equipment used and the workmanship on most residential PV systems will more closely resemble the equipment and workmanship on a commercial electrical installation than those items in a residential electrical system. There will usually be surface-mounted disconnects and much of the wiring will be in exposed, surface-mounted conduit.

The installer should have a ladder on-site the day of the inspection to facilitate examining the installed PV array. A quick look at the PV array on the roof should verify that any exposed wiring is firmly secured to the PV modules or the mounting structure and is not dangling down where it would be subject to physical damage (photo 10.5).

If the backs of the PV modules can be closely observed, proper grounding of the modules should be checked. The hardware supplied by the module manufacturer should have been used as shown in the instruction manual delivered with the permit application. Each PV module must be grounded, and if exposed, single-conductor cables touch the mounting racks or a metal roof, those objects should also be grounded. See Chapter 3 for more details on module grounding.

The conductors used for module interconnections should be as specified in the permit application with respect to size (AWG), insulation type, and temperature rating. Any PV combiners containing overcurrent devices exposed to sunlight should be noted and the plans and technical data reviewed to determine if adequate temperature deratings were applied. Conduits in sunlight will also be exposed to higher-than-ambient temperatures.

Inspect in 15 Minutes?

Yes, it *might* be possible to perform the above inspections in 15 minutes if the inspector has spent some time at the plan-check stage and is experienced in PV systems employing this inverter and the installer is there to answer questions, open the inverter and other equipment as necessary and to provide a ladder for roof access. However, any problems found in the above areas should warrant a closer look at the entire system and when more details are examined, the inspection time can grow. A lack of familiarity with either PV in general, the equipment being installed, or the installer would normally dictate that the inspection takes more time. How much? Some residential PV inspections for new inspectors are somewhat of a training session and with a knowledgeable installer, examining and discussing all the details relating to a durable, safe (for 50-years) installation might take two or more hours.

Should We Do 15-Minute Inspections?

See the little girl in photo 10.6? That PV system she is touching will still be producing power when her grandchildren are her age. It *will* take more than a 15-minute inspection to ensure that the PV system will be as safe then as it is now. Fifteen minutes is probably insufficient time to ensure the public safety of a system that may operate, with possibly little attention, over a 40–50-year period.

Photo 10.6 • A 40 kW PV system. Safe now, but will it still be safe in 40 years" *Courtesy James Worden*

A

2014 and 2017 NEC Photovoltaic Electrical Power Systems Inspector/Installer Checklist

The following checklist is an outline of the general requirements found in the *2014-2017 National Electrical Code* (*NEC*) in Articles 690 and 705 that deal with Photovoltaic (PV) Power Systems installations.

The checklist is only a guide and applies to any component used or installed in a PV system other than devices inside a listed, factory-assembled component.

The local authority having jurisdiction (AHJ) or inspector has the final say on what is or is not acceptable. Local codes may modify the requirements of the *NEC*.

This list should be used in conjunction with Article 690, Article 705 and other applicable articles of the *NEC* and includes inspection requirements for both stand-alone PV systems (with and without batteries) and utility-interactive PV systems. Where Article 690 or 705 differ from other articles of the *NEC*, Article 690 or 705 takes precedence. (690.3, 705.3)

NOTE: The *2017 NEC* has many detailed changes in Art 690 and 705 with older material being placed in new articles and numerous small revisions of the remaining material. *Code* references will generally be to the 2017 *NEC*.

CHECKLIST FOR PHOTOVOLTAIC POWER SYSTEM INSTALLATIONS

1. PV ARRAYS

☐ PV modules listed to UL Standard 1703? [110.3, 690.4(B)]

a. Mechanical Attachment

☐ Modules attached to the mounting structure per the manufacturer's instructions? [110.3(B)]

☐ Roof penetrations secure and weather tight? (110.12, 110.13)

b. Grounding

☐ Each module grounded using the supplied hardware, the grounding point identified on the module and the manufacturer's instructions? Note: Bolting the module to a "grounded" structure usually will not meet *NEC* requirements [110.3(B)] and may not comply with the instructions for grounding the PV module. Array PV mounting racks are usually not identified as equipment-grounding conductors, unless certified/listed to UL Standard 2703. (690.43) Module instruction manual must specifically show/indicate grounding and mounting method.

☐ Properly sized equipment-grounding conductors routed with the circuit conductors?(690.45)

c. Conductors

☐ Conductor type?—If exposed: USE-2 or PV wire for grounded PV arrays and PV wire for ungrounded PV arrays. All PV modules will use PV wire. 2017 *NEC* allows USE-2 or PV wire for both grounded and ungrounded systems.

☐ Conductor insulation rated at 90°C (UL-1703) to allow for operation at 70°C+ near modules and in conduit or cables exposed to sunlight? [Table 310.15(B)(3)(c)]

☐ Temperature-corrected ampacity calculations based on 125% of short-circuit current (Isc) or the 156% Isc without conditions of use (take the worst case)?

Note: *Suggest* temperature derating factors of 65°C for conductors behind modules in installations where the backs of the module receive cooling air (4" or more from roof) and 75°C where no cooling air can get to the backs of the modules. Ambient temperatures (near and at the array location) more than 40°C may require different derating factors.

☐ Portable power cords allowed only for tracker connections? [690.31(C), 400.3,7,8]

Strain reliefs/cable clamps or conduit used on all cables and cords? (300.4, 400.10)

☐ Listed for the application and the environment? Fine stranded, flexible conductor conductors properly terminated with terminals listed for such conductors? (110.14)

☐ Cables and flexible conduits installed and properly marked? (690.31)

☐ Exposed conductors in readily accessible areas in a raceway or guarded if over 30 volts? [690.31(A)] Note: Raceways cannot be connected to most modules. Conductors should be installed so that they are not readily accessible (i.e., guarded).

2. OVERCURRENT PROTECTION

☐ Overcurrent devices in the dc circuits listed for dc operation? If device not marked dc, verify dc listing with manufacturer. Auto, marine, and telecom devices are not acceptable.

☐ In PV circuits, OCPD must be listed as PV device [690.9(B)].

☐ Rated at 1.25 x 1.25 = 1.56 times short-circuit current from modules? (UL-1703, 690.8, module instructions). Overcurrent devices listed for PV applications are required. [690.9(B)].

☐ Each module or series string of modules have an overcurrent device protecting the module? [UL-1703 / *NEC* 110.3(B)]. Note: Frequently, installers ignore this requirement marked on the back of modules.

Listed combiner PV combiner boxes meeting this requirement are available. One or two strings of modules generally do not require overcurrent devices, but three strings or more in parallel will usually require an overcurrent device. The module maximum series fuse must be at least 1.56 I_{SC}. [690.9(A)]

2017 *NEC*: Only one conductor of an ungrounded PV source or PV output circuit is permitted to have an OCPD. If used, other OCPD in the dc circuits must be in the same polarity conductor.

☐ Located in a position in the circuit to protect the module conductors from backfed currents from parallel module circuits or from the charge controller or battery? [690.9(A)]

☐ Smallest conductor used to wire modules protected? Sources of overcurrent are parallel connected modules, batteries, and ac backfeed through inverters. [690.9(A)]

☐ User-accessible fuses in "touch-safe" holders or fuses capable of being changed without touching live contacts? Disconnects from all sources of voltage in dc combiners at the inverter? (690.16)

3. ELECTRICAL CONNECTIONS

☐ Pressure terminals tightened to the recommended torque specification? [110.3(B), 110.14]

☐ Crimp-on terminals listed and installed with listed crimping tools by the same manufacturer? [110.3(B)]

☐ Twist-on wire connectors listed for the environment (i.e., dry, damp, wet, or direct burial)and installed per the manufacturer's instructions?

☐ Pressure lugs or other terminals listed for the environment? (i.e., inside, outside, wet, direct burial)

☐ Power distribution blocks *listed* and not just UL Recognized?

☐ Terminals containing more than one conductor listed for multiple conductors?

☐ Connectors or terminals using flexible, fine-stranded conductors listed for use with such conductors? [690.31(H), 690.74(A), 110.14]

☐ Locking (tool-required) connectors on readily accessible PV conductors operating over 30 volts? [690.33(C)]

4. CHARGE CONTROLLERS 2017: See Article 706, Energy Storage Systems

☐ Charge controller listed to UL Standard 1741? [690.4(B)] Exposed energized terminals not readily accessible?

☐ Does a diversion controller have an independent backup control method? [706.23(B)(1)]

5. DISCONNECTS 2017 *NEC*: Substantial changes

☐ Rapid shutdown system installed? (690.12) Applied to inverter inputs, module outputs, batteries (where used), and combiner outputs? Listed equipment is available and the UL Standards addressing the requirements are published. Operation verified?

☐ Disconnects listed for dc operation in dc circuits? Automotive, marine, and telecom devices are not acceptable.

☐ PV Disconnect readily accessible and located at first point of penetration of PV conductors? Location not specified in the 2017 *NEC*.

☐ PV conductors outside structure until reaching first readily accessible disconnect unless in metallic raceway? [690.13(A), 690.31(F)] Metallic raceway now required all the way to the inverter dc input.

☐ Disconnects for all current-carrying conductors for the PV system? (690.13) 2017 *NEC*: PV System Disconnect must disconnect all circuit conductors—even on solidly grounded systems -but solidly grounded conductors should not be opened. (Look for a possible change to this requirement in the 2020 *NEC*.)

☐ Disconnects for equipment? (690.15/690.17) 2017 *NEC*: Equipment isolation disconnects may disconnect only the ungrounded conductor.

☐ DC combiner has output circuit disconnect/isolator internal or within 3 m (10 ft)? [690.15(A)]

☐ Grounded conductors *not* fused or switched (except PV system disconnect in *2017 NEC*)?

Note: Listed PV Power Centers are available for 12, 24, and 48-volt systems and they contain charge controllers, disconnects, and overcurrent protection for entire dc system with possible exception of source circuit or module protective fuses.

6. INVERTERS (Stand-alone Systems)

☐ Inverter listed to UL Standard 1741? [110.3, 690.4(B)] Note: Inverters listed to telecommunications, automotive or other standards do not meet *NEC* requirements.

☐ DC input currents from the battery calculated for conductor and fuse requirements? Input current = rated ac output in watts divided by lowest battery voltage divided by inverter efficiency at that power level. [690.8(A)(4)]

☐ Cables to batteries sized 125% of calculated inverter input currents? [690.8(A), 706.20]

☐ Overcurrent/Disconnects mounted near batteries and external to PV load centers if conductors are longer than 4–5 feet to batteries or inverter?

☐ High interrupt, listed, dc-rated fuses or circuit breakers used in battery circuits? AIR/AIC at least 20,000 amps? [706.21 110.9, 110.10]

☐ No multi-wire branch circuits where single 120-volt inverters are connected to 120/240-volt load centers? [Art 100–Branch Circuit, Multi-wire), 710.15]

7. BATTERIES

For *2017 NEC*: See Article 706, Energy Storage Systems

☐ No separate batteries cells are listed.

☐ AC Battery systems are generally self-contained and will be listed as an assembly.

☐ Building-wire type cables used? (Chapter 3)

Note: Welding cables, marine, locomotive (DLO), appliance wire material (AWM) and auto battery cables don't meet *NEC* requirements. Flexible, listed RHW, or THW cables are available. Article 400 flexible cables larger than 2/0 AWG are OK for battery cell connections, but not in conduit or through walls [690.74, 400.8]. Flexible, fine-stranded cables require limited-availability, specially-listed terminals [110.14, 690.74]. See stand-alone inverters for ampacity calculations.

☐ Access limited? [706.30]

☐ Installed in well-vented areas (garages, basements, outbuildings, and not living areas)? Note: Manifolds, power venting, and single exterior vents to the outside are not required and should be avoided [706.10(A)].

☐ Have the conductor routing and protection requirements of 706.20 and 706.32 been met? Cables to inverters, dc load centers, and/or charge controllers in conduit?

☐ Conduit enters the battery enclosure below the tops of the flooded batteries? (300.4)

Note: There are few listed battery boxes. Lockable heavy-duty plastic polyethylene tool boxes are usually acceptable.

8. INVERTERS (Utility - interactive Systems)

☐ Inverter listed to UL Standard 1741 and identified for use in interactive photovoltaic power systems? [690.4(B), 705.4]. Note: Inverters listed to telecommunications and other standards do not meet *NEC* requirements.

☐ Backup charge controller to regulate the batteries in systems with multimode inverters when the grid fails? [706.23(B)]

☐ Connected to dedicated branch circuit with back-fed overcurrent protection? [705.12(B)] or connected as a supply-side connection with overcurrent protection within 10 feet? [705.12(A), 705.31]

☐ Listed dc and ac disconnects and overcurrent protection? (690.15, 17)

☐ All requirements of 705.12(B) or 705.12(A) met?

NOTE: Square wave or modified sine wave inverters may be listed to UL 1741 but are not compatible with many power tool battery chargers, smoke alarms, and other listed electronic devices and should not be used with these devices. The manufacturer's instruction manual will usually have the warning statement [110.3(B)].

9. GROUNDING

NOTE: Grounding and Disconnects extensively revised in *2017 NEC.*

☐ Only one bonding conductor (grounded conductor to ground-only on solidly grounded PV systems) for dc circuits on grounded PV arrays and one bonding conductor for ac circuits (neutral to ground) for ac system grounding?

Note: The utility-interactive inverter will generally provide the functional ground for the system. Instructions for that functional ground will be in the inverter installation manual.

- ☐ System/inverter grounding meets requirements of 690.47?

- ☐ Equipment grounding conductors properly sized (even on ungrounded, low-voltage systems)? (690.43, 45, 46)

- ☐ Disconnects and overcurrent in both ungrounded conductors in each circuit on 12-volt, ungrounded systems or on ungrounded systems at any voltage? (690.9, 690.13, 690.15, 690.31) Revised for *2017 NEC* with functional grounding. Note that functionally grounded PV systems under the 2017 *NEC* will not have any dc PV source or dc PV output conductors with white insulation.

- ☐ Bonding-grounding fittings or bushings used with metal conduits when dc system voltage is more than 250V dc? (250.97). Grounding bonding bushings used where grounding electrode conductors are in metallic raceways and /or enclosures?

10. CONDUCTORS (General)

- ☐ Standard building-wire cables and wiring methods used? [300.1(A)]

- ☐ Wet-rated conductors used in conduits in exposed locations? (100 Definition of Location, Wet)

- ☐ Conductor insulations other than black in color will not be as durable as black in the outdoor UV-rich environment.

- ☐ DC color codes correct? They are the same as ac color codes—grounded conductors are white or gray and equipment-grounding conductors are green, green/yellow, or bare. [200.6(A)]

- ☐ Ungrounded PV array conductors on ungrounded PV arrays will *not* be white in color. Note that functionally grounded PV systems under the 2017 *NEC* will not have any dc PV source or dc PV output conductors with white insulation. The color white will only be used in solidly grounded PV systems (690.41).

11. Markings

- ☐ All field-applied markings correct? [690.13(B), 690.31(B), 690.51, 690.53, 690.54, 690.55, 705.10, 705.12]

- ☐ Meet color and letter size requirements? (690.56)

12. DC PV Arc Fault Circuit Protection?

- ☐ Usually installed in the inverter or on larger systems in the array field. May be multiple devices (690.11). UL Standard 1699B is the applicable standard.

13. PV Rapid Shutdown Systems?

- ☐ Installed per 690.12 and local requirements? Operational?

B

Bibliography

Molded-Case Circuit Breakers, Molded-Case Switches, and Circuit-Breaker Enclosures for Use with Photovoltaic (PV) Systems. UL Standard 489B, 2016.

National Fire Protection Association. *National Electrical Code® (NEC®) Handbook.* Quincy, Massachusetts: NFPA, 2017.

National Fire Protection Association. *National Electrical Code® (NEC®) Handbook.* Quincy, Massachusetts: NFPA, 2014.

National Fire Protection Association. *NFPA 70: National Electrical Code.* Quincy, Massachusetts: NFPA, 2017.

National Fire Protection Association. *NFPA 70: National Electrical Code.* Quincy, Massachusetts: NFPA, 2014.

Photovoltaic (PV) DC Arc-Fault Circuit Protection. UL Standard 1699B, 2011.

Solar American Board of Codes and Standards. *Expedited Permit Process Report - Revision 2*, last accessed on 12/13/2017. http://solarabcs.org/about/publications/reports/expedited-permit/index.html.

Standard for Connectors for Use in Photovoltaic Systems. UL Standard 6703, 2014.

Standard for Flat-Plate Photovoltaic Modules and Panels. UL Standard 1703, 2012.

Standard for Flat-Plate Photovoltaic Modules and Panels. UL Standard 1703, 2002.

Standard for Industrial Control Equipment. UL Standard 508, 1999.

Standard for Inverters, Converters, Controllers and Interconnection System Equipment for Use With Distributed Energy Resources. UL Standard 1741, 2010.

Standard for Mounting Systems, Mounting Devices, Clamping/Retention Devices, and Ground Lugs for Use with Flat-Plate Photovoltaic Modules and Panels. UL Standard 2703, 2015.

Standard for Photovoltaic Wire. UL Standard 4703, 2011.

Wiles, John, "Perspectives on PV archives," IAEI Magazine, 2015-2018.

Photovoltaic Power Systems
For Inspectors, Plan Reviewers & PV Professionals

BASED ON THE 2017 *NATIONAL ELECTRICAL CODE*

Third Edition

Author John Wiles

Made in the USA
Monee, IL
02 November 2026

67383027R00109